W9-BUL-486

READ TO WORK

TECHNOLOGY

WENDY HARRIS

Author: **Wendy Harris**
Series Editorial Consultant: Harriet Diamond, *President, Diamond Associates, Multifaceted Training and Development, Westfield NJ*

Director, Editorial & Marketing, Adult Education: Diane Galen
Market Manager: Will Jarred
Assistant Market Manager: Donna Frasco
Editorial Development: Learning Unlimited, Inc.
Project Editors: Douglas Falk, Elena Petron
Editorial Assistant: Kathleen Kennedy
Production Director: Kurt Scherwatzky
Production Editor: John Roberts
Art Direction: Pat Smythe, Kenny Beck
Cover Art: Jim Finlayson
Interior Design & Electronic Page Production: Levavi & Levavi
Photo Research: Jenifer Hixson

Photo Credits: p. 6: Amy C. Etra, Photo Edit; p. 14: Michael Newman, Photo Edit; p. 22: Michael Newman, Photo Edit; p. 32: David J. Sams, Stock Boston; p. 40: Matthew Borkowki, Stock Boston; p. 48: Tony Freeman, Photo Edit; p. 58: Cindy Charles, Photo Edit; p. 66: Tony Freeman, Photo Edit; p. 74: Kathy Ferguson, Photo Edit; p. 84: John Coletti, Stock Boston; p. 91: M. Dwyer, Stock Boston; p. 100: Dean Abramson, Stock Boston.

Copyright © 1997 by Pearson Education, Inc., publishing as Pearson Learning Group, 299 Jefferson Road, Parsippany, NJ 07054. All rights reserved. No part of this book may be reproduced or transmitted in any form or by any means, electronic or mechanical, including photocopying, recording, or by any information storage and retrieval system, without permission in writing from the publisher. For information regarding permission(s), write to Rights and Permissions Department.
ISBN 0-8359-4688-6
Printed in the United States of America
3 4 5 6 7 8 9 10   05 04 03 02

1-800-321-3106
www.pearsonlearning.com

C O N T E N T S

TO THE LEARNER

Welcome to the *Read To Work* series. The books in this series were written with you, the adult learner, in mind. Good reading skills are important in the world of work for these reasons:

◆ They may help you get the job you want.
◆ They will help you learn how to do your job well.
◆ They can help you get a better job.

The lessons in this book, *Read To Work: Technology,* will help you improve your reading skills. As you work through the lessons, you will also learn about jobs in electronics, computers, communications, and environmental technology.

UNITS

Read To Work: Technology is divided into four units. Each unit covers different kinds of jobs. You can look at the **Contents** to see what fields and jobs are covered in this book.

LESSONS

Each unit contains at least 3 lessons. Each lesson teaches one reading skill and covers one kind of job. Here are some things to look for as you read each lesson:

Words to Know are words you will learn in the lessons. Look for the meaning of each new word to the left of what you are reading. You will also see a respelling of the words like this: *pronunciation* (proh-nun-see-AY-shuhn). This respelling will help you say the word correctly. There is a guide to help you with the respellings on page 105.

Job Focus describes the job in the lesson. It also tells you what types of skills are needed to do the job.

How It Works teaches you about the reading skill and how you can use it.

Readings include memos, pages from handbooks and manuals, posters, product guidelines, safety notices, and articles from company newsletters. If you look through this book, you will see that the reading passages look different from the rest of the lesson. They are examples of reading materials from the world of work.

Check Your Understanding questions can be multiple choice, short answer, or true/false. They will help you check that you understand the reading.

On the Job gives you a chance to read about real people as they do their jobs.

OTHER LEARNING AIDS

There are other learning aids at the back of the book. They are:

Respelling Guide: help with pronouncing words
Resources: where to get more information on the jobs in the book
Glossary: definitions of the Words to Know
Index: job names in the book
Answer Key: answers to *Check Your Understanding* and *Lesson Wrap-Up* questions

Now you are ready to begin using *Read To Work: Technology*. We hope that you will enjoy this book and learn from it.

Laser Technology Occupations

Workers in laser technology use or build machines that have lasers. Lasers are steady, powerful beams of light that are used in production and in health care. Some assemblers build the machines that use lasers, while other assemblers build the lasers themselves. Laser technicians set up the machines that have the lasers. Laser operators run the laser equipment.

Laser technology workers read many different types of materials, including manuals, safety instructions, and memos. They also need to understand pictures, charts, and other types of diagrams that help them to find information quickly.

This unit teaches the following reading skills:

◆ finding the stated main idea
◆ finding details that support the main idea
◆ understanding visual information, such as charts and graphs.

You will learn how workers in laser technology jobs use these skills in their work.

L e s s o n I

Working as an Assembler

▼▼▼▼▼▼▼▼▼▼▼▼

Words to Know

assemblers

continuing education

human resources

laser

precision lenses

scalpels

scanners

sensors

supervisor

Most jobs in technology require workers to keep learning on the job. Sometimes, workers have to learn new skills and information. They need to read company manuals, memos, or textbooks. They may read articles in magazines or newsletters. They may take a specific class or course. No matter how or when workers read new material, they must be able to understand it.

A key to understanding what you read is learning to find the writer's main point. This is called **finding the main idea**. This skill will help you understand the most important ideas in what you read.

Have you ever been asked, "What was the book about?" Has a friend ever asked you, "What was the movie about?" Next time this happens, try answering with a single sentence. That sentence will be your main idea.

Job Focus

In this lesson, you will learn about **electro-mechanical assemblers,** also called EMAs. EMAs put together machines that contain electronic parts. EMAs build the machines that use lasers to make other goods. However, EMAs do not build the actual lasers.

Electro-mechanical assemblers must be able to tell good products from bad ones. They must be able to judge the quality of their own work. EMAs often attach wires to a connector. They must be able to tell that the wires are placed correctly, that the right amount of wire is showing, and that the wires are being held in the right place.

EMAs work with small machine parts. They must be skillful with their hands, and they must also have very good eyesight.

2

Finding the Main Idea: How It Works

The main idea is the most important idea in your reading. **Finding the main idea** helps you sum up what reading material is about. Every reading has an overall main idea.

Each paragraph will also have its own main idea. Often, the main idea is stated in the first or last sentence of a paragraph. The main idea can also appear in the middle of a paragraph. Sometimes, it is not directly stated. You will learn how to figure out an unstated main idea in Lesson 11. To find a stated main idea, follow these steps:

1. **Identify the topic.** Try to find a single word or a few words that clearly describe what the reading is about.
2. **Ask what the reading says about the topic.** Try to sum up the important idea. Look for the most important idea in each paragraph.

Read the memo below, and think about the main idea.

assemblers
(uh-SEHM-blerz) workers who put together equipment

human resources
(HYOO-muhn REE-sawrs-ehz) department in a company responsible for worker hiring and benefits

continuing education
(kuhn-TIHN-yoo-ihng ehj-yoo-KAY-shun) classes to help adults on the job and in their personal lives

laser (LAY-zer) machine that produces a steady, powerful stream of light

supervisor (SOO-per-veyez-er) person who manages other workers

M E M O

Date: April 19
To: **All Assemblers**
From: **Human Resources**
Subject: **Continuing Education** Opportunities

Here is important information about new courses for workers. This June we are offering four **laser** courses.

The four courses are: Laser Basics, Laser Safety, Introduction to Industrial Lasers, and Introduction to Medical Lasers. There is room for up to 8 people per class.

Each class will be offered two times during the week of June 10th. Classes will be held on Monday and Thursday. The first time period is 8:00 a.m.-10:00 a.m. The second time period is 7:00 p.m.-9:00 p.m.

Please select the time that is best for your shift. You must clear your selections with your **supervisor.**

First, underline the topic, or subject, of the memo. Write it on the line below.

_____

If you wrote *continuing education opportunities,* you are right. It is stated at the top of the memo.

Next, find an idea that sums up what the whole memo is about. Write it here.

_____

If you wrote: *Here is important information about new courses,* you are right. It is stated in the first paragraph.

precision lenses
(pree-SIHZJ-uhn LEHNZ-ehz)
high-quality, carefully shaped
pieces of curved glass

The following selection is a sample page from the Laser Basics course manual. As you read it, identify the topic. Then, think about the main idea.

COURSE MANUAL FOR LASER BASICS

This manual will help you learn about lasers. There are key ideas and facts that you need to know to be able to work with lasers. These laser basics are the key to understanding how lasers work and how they are used.

I. INTRODUCTION TO LASERS

1.1 A **laser** is used to create a special type of light. The word *laser* is made up from the first letter of a series of words. The letters in the word laser stand for **L**ight **A**mplification (AM-pluh-fih-KAY-shuhn) by **S**timulated (STIHM-yoo-lay-tuhd) **E**mission (ih-MISH-uhn) of **R**adiation (RAY-dee-AY-shuhn). Light from a laser is different from ordinary light, such as the light from a flashlight.

1.2 The diagrams below show how light produced by a laser differs from light produced by more ordinary sources.

1.3 Ordinary light sources, as in Figure A, let light out across a wide area. This allows the light to travel in many different directions. This also means that the light beams are not focused.

1.4 Lasers strengthen beams of light and make them more powerful. **Precision lenses** and other parts of the laser bounce the light back and forth in the laser chamber, Figure B. The lens at one end of the chamber only bounces light back. The other end allows a focused beam of light to shine through.

1.5 All of the light waves produced by a laser are the same. They have the same length *(wavelength)*. The waves also have the same *frequency*. This means that the number of waves that pass a point in one second is the same. The result is that lasers are more powerful and focused than other light sources.

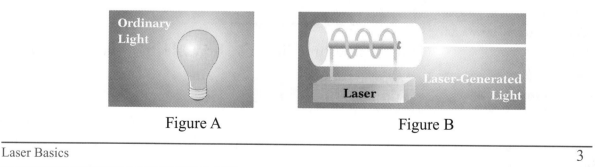

Figure A   Figure B

Answer each question based on the reading on page 4.

1. What is the topic of the page from the course manual?

   a. light
   b. EMAs
   c. lasers
   d. lenses

2. What is the main idea of the whole page?

   a. The word laser is based on a series of words that start with the letters L-A-S-E-R.
   b. There are certain facts that you need to know to be able to work with laser equipment.
   c. Light produced from a laser is different from ordinary light.
   d. Laser-made light has special qualities that make lasers powerful and exact.

3. What is the main idea of paragraph 1.3?

   a. Laser light is produced by a flashlight.
   b. Ordinary light sources let light out across a wide area.
   c. Light waves produced by a laser are all the same.
   d. Light beams from ordinary sources are not focused.

4. What is the main idea of paragraph 1.4?

   a. Lasers make beams of light more powerful.
   b. Precision lenses and other parts of the laser bounce the light back and forth in the laser chamber.
   c. The lens at one end of the chamber bounces light.
   d. The other end of the chamber allows a focused beam of light to shine through.

5. What is the main idea of paragraph 1.5?

   a. The same number of laser light waves pass through a given point every second.
   b. Light waves from lasers have the same length.
   c. Light waves from lasers have the same frequency.
   d. All the light waves produced by a laser are the same.

Check your answers on page 113.

Simon begins his day very early. Simon is an EMA at LaserVu. He works the 7:00 A.M. to 3:00 P.M. shift.

Simon came to LaserVu without training in lasers. However, he had always enjoyed math, science, and working with his hands. He received his GED and went to work for an electronics company. There, he learned how to assemble different types of electronics equipment. He often operated a soldering iron, a tool that melts metal. His work experience and interests made LaserVu want to hire him.

But before hiring Simon, LaserVu asked him to take a reading and math skills test. The test included a section on reading simple diagrams. Simon did well on the test. His education and job experience had helped him.

At LaserVu, Simon is learning to build machines that use lasers. He works on industrial laser assembly. He operates a soldering iron, as he did at his former job. Sometimes, he reads directions and diagrams that tell how machine parts are put together. He also replaces defective parts—those parts that don't work.

Simon's supervisor saw that he was eager to work on new projects. When LaserVu decided to create a team in his department, Simon was asked to be the assembler. Simon believes his enthusiasm for his work and his talented hands got him this chance. He values this job because he gets to give his opinions and make suggestions.

Simon enjoys the variety of projects he gets to work on at LaserVu. He hopes to get more training someday so that he can work on medical laser equipment.

TALK ABOUT IT

1. Explain why LaserVu decided to train Simon to be the EMA on the team. What skills and qualities did he have?

2. Discuss what is most interesting to you about Simon's job.

Company newsletters contain stories and articles of interest for workers at all levels. Companies are always developing new products and services. They use their newsletters to educate workers about them.

Newsletters also tell workers about new projects that the company is planning. Read the following page from LaserVu's newsletter.

LaserVu, Inc.

Looking to the Future

It can be used to cut metal or treat the most delicate parts of our bodies. What is it? You work with it every day—it's laser technology.

From the time the first laser was built in 1960, hundreds of different lasers have been built and used. Whether the laser equipment you build is for industrial or medical purposes, LaserVu builds whatever our customers need. LaserVu is proud to be on the forefront of laser technology.

Medical Laser Equipment from LaserVu

LaserVu employees have built and serviced a wide range of medical lasers. These lasers are taking the place of **scalpels** and other cutting instruments. Patients heal faster. Their bodies have fewer scars. Doctors use our lasers to perform the most delicate surgeries. Our medical lasers are on the cutting edge of today's technology.

Industrial Laser Equipment from LaserVu

Laser cutters, laser welders, **scanners**, and laser **sensors** are just some examples of laser equipment assembled, delivered, and serviced by you and your co-workers at LaserVu. We've supplied dozens of laser devices to the automotive, electronics, technology and graphic arts industries.

---

scalpels (SKALP-ehlz) small knives with sharp blades used in surgery

scanners (SKAN-erz) machines that use beams of light to read type and images by comparing dark and light spaces

sensors (SEHN-serz) parts of equipment that pick up, or "sense," messages or information

Answer each question based on the newsletter on page 7.

1. In which paragraph is the main idea of the entire article stated?

  a. paragraph 1
  b. paragraph 2
  c. paragraph 3
  d. paragraph 4

2. Which of the following statements best sums up the main idea of the article?

  a. Lasers are used for medical purposes.
  b. Lasers are used for industrial purposes.
  c. LaserVu produces a wide variety of laser equipment.
  d. LaserVu is proud to be a leader in laser technology.

3. What is the main idea of paragraph three?

  a. LaserVu builds and services a wide range of laser medical equipment.
  b. LaserVu lasers help patients heal more quickly.
  c. LaserVu lasers are taking the place of scalpels.
  d. LaserVu lasers are used by doctors to perform surgery.

4. What is the topic of paragraph four?

  a. laser sensors
  b. medical lasers
  c. industrial lasers
  d. laser technology

5. Imagine you are a LaserVu salesperson. You need to tell a customer about industrial lasers. You decide to use the main idea from the paragraph on industrial lasers in the newsletter. Below, write the main idea of paragraph four.

Check your answers on page 113.

◆ LESSON WRAP-UP

In this lesson, you learned how to:
- find the stated main idea of an entire reading
- find the stated main idea of a paragraph

The main idea is the most important idea. Finding the main idea helps you fully understand what you read. You learned two steps for finding the main idea:

1. Identify the topic.
2. Look for the most important idea.

Your employer may ask you to read different materials on the job. You may be asked to read a memo, booklet, newsletter, or report. Finding main ideas when you read on the job will help you understand the materials.

Think about something that you read recently. It may have been something for your job. Or it may have been an article from a magazine or newspaper. If possible, bring it in to class.

1. State the name of the reading.
2. State the topic of the reading.
3. State the main idea of the reading.
4. Explain how finding the main idea helped you understand the reading.

1. I read

2. The topic of the reading is

3. The main idea of the reading is

4. Understanding the main idea of the reading helped me

Check your answers on page 113.

Working as a Laser Technician

Words to Know

bunny suits

contamination

particles

policies

termination

violation

visor

Employers care about worker safety. They want to make sure that workers are protected on the job. To do this, they often post, or display, signs with general safety rules. They may also review the rules with the workers.

Suppose you are a laser technician. You are reading a safety poster on the employee bulletin board. At the top it says, *"It is important to follow all the safety guidelines listed below."* You know that this is the poster's main idea, or most important point. And you would expect an explanation to follow.

When you read material on the job, you need to **find supporting details** as well as the main idea. The supporting details explain and tell more about the main idea. On the poster, the safety rules and procedures would be the supporting details.

Job Focus

In this lesson, you will learn about the job of **laser technician** (tehk-NIHSH-uhn). Laser technicians work in industry and health care. They make sure that lasers operate correctly. They check for problems with the lasers. This is called *troubleshooting*. They may also fix some of the problems that they find.

Laser technicians set up lasers to produce certain parts. To do their job, technicians may need to read diagrams. Laser technicians must be good at understanding details. Their work is very exact.

Laser technicians need good communication skills. They speak at meetings and on the phone. They need to communicate with supervisors and co-workers. They may also need to write memos, reports, and notes.

Jobs in the laser industry are expected to keep growing at a very strong rate.

Finding Supporting Details: How It Works

Supporting details are the pieces of information that help the reader understand the main idea. These details may be rules, dates, and other facts and examples. Looking for details that support the main idea is important when you read on the job.

Read the safety poster below. Look for the main idea and the supporting details.

A Safe Workplace at Laser Light

It is important to follow all the guidelines listed below. Laser Light hopes to maintain our fine safety record. Your cooperation will make sure this happens.

LASER LIGHT SAFETY GUIDELINES

Protect Your Eyes. Don't look at the laser beam. This can damage your eyes. Wear protective safety goggles or glasses in laser work areas. Put on the goggles or glasses before entering the laser work room. Wear them at all times.

Protect Your Hands. Never put your hand under the laser beam. Close the door to the laser cabinet before you turn on the laser. Make sure your fingers never touch the buffer. The wheel moves very quickly. It can cut off your fingers if they come in contact.

Protect Your Work. When handling materials, make sure you wear two layers of gloves. Put the cotton gloves on first to protect your skin. Put latex gloves over the cotton gloves. This keeps fibers from the cotton gloves away from the work and the laser.

The main idea of the poster is *It is important to follow all the safety guidelines*. Write the three details that support this main idea.

1. _____

2. _____

3. _____

You were right if you wrote: *wear goggles to protect your eyes; protect your hands by keeping them away from the laser; wear gloves to protect your work.*

Notice that a supporting detail can also be the main idea of its own paragraph. It can be supported by its own details. List the details that support *Protect Your Eyes*.

policies (PAHL-uh-sees) general practices, guidelines, or standards

violation (vy-oh-LAY-shun) breaking of a rule

termination (ter-muh-NAY-shun) ending, as in separation from a job

———————————————

———————————————

You were right if you said *Don't look at a laser beam. Wear safety goggles.*

Many employers put company rules in an employee manual. Read the page below from Laser Light's employee manual.

The Laser Light Drug-Free Workplace Policy

Our company has the goal of a drug-free workplace. To achieve this goal we have adopted the following **policies:**

☞ You will not manufacture, sell, hand out, carry, or use illegal drugs at Laser Light. Employees who do not obey this policy will be violating a major company policy. **Violation** may result in **termination** from your job.

☞ If an employee uses drugs while on the job, the following steps will be taken. The employee's supervisor will be informed. The supervisor will refer the employee to the Human Resources Manager. The Human Resources Manager will help the employee find a treatment program. The employee must keep the Human Resources Manager informed about his/her progress in the program.

☞ Laser Light requires every job applicant to be screened for illegal drug use before being hired. Laser Light will pay for this test. Every employee will be given a follow-up test within 3 months of their employment. Laser Light will cover the cost of the follow-up test as well.

☞ Laser Light offers drug-awareness and education services to its employees. We run a two-hour drug awareness program twice a year. Laser Light will pay for any employee to attend Shady Glen Hospital's drug awareness program. We also have a number of videotapes and books in our library. These materials can be checked out and used at your own home.

The company will hand out this policy statement every year. We will also post this policy on the three bulletin boards. You also have a copy in your employee manual.

LASER LIGHT MANUAL   19

Some main ideas and supporting details from "The Laser Light Drug-Free Workplace Policy" on page 12 are listed below. Match each supporting detail with its main idea by writing the letter of the supporting detail on the correct line. Then answer the remaining questions.

Main Ideas

1. If an employee uses drugs on the job, several steps are taken. _____

2. Laser Light has every employee screened for illegal drugs before being hired. _____

3. Laser Light offers its employees free drug-awareness and education services. _____

Supporting Details

    a. Videotapes and books are available.

    b. The human resources manager recommends a treatment program.

    c. Progress meetings are held with human resources.

    d. Laser Light pays for the drug testing.

    e. Two-hour drug awareness programs are held twice a year.

    f. Laser Light pays for employees to go to Shady Glen's drug awareness program.

    g. New employees are given follow-up drug tests within three months after being hired.

4. In what three ways does Laser Light inform its employees about its drug-free workplace policy?

    a.

    b.

    c.

5. What is the worst thing that might happen to an employee for violating the company's drug policy?

6. Do you think it is important for Laser Light to have a drug-free workplace? State your answer as your main idea. Then, give reasons to explain why you feel as you do. Each reason will be a supporting detail for your main idea.

Check your answers on page 113.

ON THE JOB

Aisha began working at Laser Light as a laser operator. To be promoted to laser technician, she needed to finish high school and take technical courses. She had dropped out of school over 10 years earlier. However, she went back to night school and got her high school diploma. Then, she studied laser technology at a local college. Laser Light paid for her education. Getting the promotion to laser technician made Aisha feel very proud.

Aisha's work as a laser technician requires good skills with details. She programs the computer to cut materials into various shapes using lasers. After she programs the computer, she makes a few samples. These samples show workers what the final product will look like.

Aisha checks the samples to make sure that they are exactly as they should be. If the samples are good, she gives each laser operator a sample. She explains how each part of the sample should look when it's done.

Sometimes, Aisha programs computers to weld (melt) metal pieces together. As always, she must inspect each sample for possible flaws. (*Flaws* are errors or defects in a product.)

Aisha likes to explain how something gets made. She is good at helping co-workers spot problems. Most of all, she enjoys having a responsible job.

TALK ABOUT IT

1. Explain why Aisha and other laser technicians must be good with details.

2. A laser technician also needs good communication skills. Discuss ways that Aisha showed these skills on her job.

14

LESSON 2 ◆ WORKING AS A LASER TECHNICIAN

Laser technicians work in a variety of places. Some laser technicians work with small instruments in a clean room. A *clean room* is a controlled area. The air passing through the room is also carefully controlled. The air quality must be good, and it must be clean.

The reading below shows safety rules that a laser technician reads before working in a clean room. Read the rules. Then, answer the questions that follow.

contamination
(kuhn-TAM-uh-NAY-shun)
pollution or dirt

bunny suits (BUN-ee soots)
special suits that cover and
protect the body

particles (PAHRT-ih-kuhlz)
very tiny pieces

visor (VY-zer) movable brim
attached to a hat or hood

Clean Room Safety Rules

Before you begin working in a clean room, please read the safety rules below. Keeping the room clean and free of **contamination** is important.

1.   **Wear a bunny suit.** In the dressing area outside the clean room, you will find special suits called **bunny suits.** Bunny suits prevent fibers of your own clothing from escaping onto your workbench or into the air. The suit must cover your entire body, from your neck to your ankles. You must be able to move easily in the bunny suit. Once you have selected your suit, put your name on the tag inside.

2.   **Cover your head.** All employees must cover their heads with hoods. Hoods help keep small **particles** of dust or skin from falling on the lenses or laser. Even bald employees must cover their heads. Any employee with long hair should tie it back before putting on the hood. Make sure all your hair is tucked in the hood.

3.   **Wear special slippers.** In the dressing area outside the clean room, you will find special slippers. Pick a pair that fits easily over your shoes. Closed-toe shoes are required under slippers. Make sure to place the top of the slippers under the ankle trim on the bunny suit.

4.   **Wear protective eyewear.** You must protect your eyes in the clean room. A clear **visor** can be attached to the hood of your bunny suit. Or, you may wear your safety goggles. Be sure to clean them before you enter the clean room.

Finish each sentence below with details from the "Clean Room Safety Rules" on page 15.

1. You can find special clean-room clothing in the

2. You will need to put on a   ,
                                    and
before entering the clean room.

3. Three things that must be covered before entering a clean room are

and

4. Clean room rules prevent contamination by:

and

5. You can use   or

to protect your eyes in the clean room.

Read each statement below. Decide if each statement is true or false. Write **T** for true and **F** for false.

_____ 6. Bald employees do not need to wear hoods.

_____ 7. Very small instruments control clean rooms.

_____ 8. Clean rooms should not be contaminated.

_____ 9. Laser technicians dress inside the clean room.

Write your answer to each question in the space below it.

10. Describe how to select the right bunny suit.

11. Describe what you think it might be like to work in a clean room. Use specific details.

Check your answers on page 113.

◆ LESSON WRAP-UP

This lesson taught you how to find supporting details. You learned that supporting details are pieces of information that help the reader understand the main idea. You also learned how to connect details with a main idea.

Employees often need to read and check details while on the job. Employees may read details in employee manuals, checklists, safety posters, and many other types of materials.

Think about how you can use supporting details in your life to support, or explain, a main idea.

Have you ever had to write an apology to someone? In the space below, write an apology note. Practice using supporting details to explain your apology.

- In the first sentence, explain what you are sorry for.
- In the next paragraph, give details about what happened.
- In the last paragraph, explain what you will change or why the situation won't happen again.

Dear   ,
I am sorry about   that happened .
First, let me explain that

I hope that

Sincerely,
(Your Name)

Check your answers on page 113.

Working as a Laser Operator

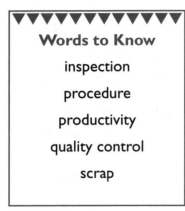

▼▼▼▼▼▼▼▼▼▼▼▼
Words to Know

inspection

procedure

productivity

quality control

scrap

Information can be shared at work in different ways. Many people read manuals, letters, or memos on the job. These types of materials use words (text) to explain ideas.

Materials on the job also include drawings, charts, graphs, or maps. These types of materials are called *visuals*. They show information with pictures, lines, bars, and other shapes.

Visual information is very useful. Diagrams can show how to put something together. Drawings can show how a product should look when it is made correctly. Graphs compare the quality of products that companies produce. **Understanding visual information** is important.

Have you ever used a diagram to put together a child's bicycle or toy? Think about a time when you read instructions that included pictures or diagrams. Did the visuals help you understand the text?

Job Focus

Laser operators run the machines that use lasers to make different materials. Laser operators may run one or more machines.

Laser operators use lasers to shape, cut, and weld different materials, such as plastics or metals. Sometimes, they work with ceramics (suh-RAM-ihks), or materials made from clay. They may use one laser to shape the edges of ceramic plates. They may use another laser to cut tiny holes in the same plate. They may use other equipment to smooth and finish their work.

Lasers are used to make more and more products every year. Many businesses call upon laser shops to produce the products that they need. The job outlook for laser operators is good.

Understanding Visual Information: How It Works

Understanding visual information is useful on the job. Business information is often reported in charts and graphs. Take a look at the two visuals below. They both report the same information—the amount of overtime that workers in the ceramics and plastics area worked during the first week in May.

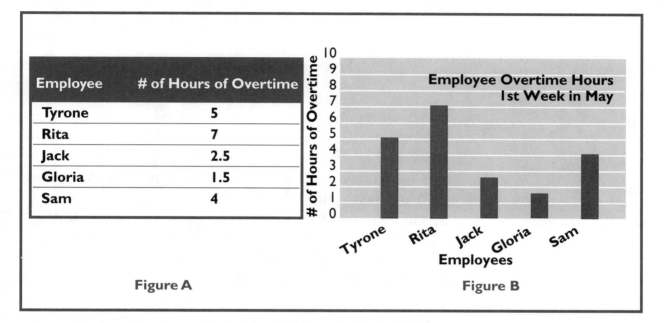

Employee	# of Hours of Overtime
Tyrone	5
Rita	7
Jack	2.5
Gloria	1.5
Sam	4

Figure A

Figure B

Tables report data (information in numbers) using columns, rows, and cells. Information in columns is arranged in a vertical, or up and down, order. The table in Figure A has two columns. In this table, one column includes *employees' names*. The other column shows *hours of overtime* worked.

Rows are used to arrange information horizontally, or left to right. Each employee has his or her own row in the table.

Cells show specific information where a row and a column meet. On the table, look at the amount of overtime Jack worked. First, find Jack's *row*. Then find the *column* for hours of overtime. Place a finger on your left hand at the beginning of Jack's row. Place a finger on your right hand on the overtime hours column. Move your left finger across and your right finger down. They should meet in the cell containing Jack's overtime hours.

In Figure A, how many hours of overtime did Jack work during the first week of May? _____

You are correct if you said *2.5 hours*.

You can also read data in *graphs*. Most bar graphs have a horizontal line across the bottom. This line runs left to right. They also have a vertical, or up and down, line. Each piece of information is plotted, or set, at a point, and a bar is drawn to it.

In the bar graph (Figure B), move your right finger along the horizontal line until you find the bar that shows Sam's overtime hours. Place your finger at the top of the bar. Move it back to the left until it hits the vertical line.

In Figure B, how many overtime hours did Sam work?

_____ You were right if you wrote *4 hours*.

The company wants to know how many good products or parts are made. It also wants to know how many parts are bad and need to be thrown away as scrap.

Companies use quality control information to improve their products and to keep their costs down.

Here are a memo and a table that a laser operator might use on the job. The memo tells how to fill out the table. Read the memo and look over the table. Then, answer the questions that follow.

quality control
(KWAWL-ih-tee kuhn-TROHL)
method used by companies to produce better products

inspection (ihn-SPEHK-shun)
checking of parts to see if they are good parts or scraps

scrap (skrap) damaged parts; discarded products

M • E • M • O

To: All Laser Operators
From: **Quality Control** Coordinators

All laser operators must track every part you receive. You must report each part that you work on. Floor supervisors will collect your forms and combine them into a master form, like the one that follows.

Please fill in each cell for each job you work on. You should report on the number of total parts you receive in each batch. You should report the number of good parts you produce that pass **inspection.** You must also track how much **scrap** you produce. Also, please chart how much time you spend on each job.

Employee	Laser #	Parts Received (in Batch)	Good Parts Produced	Scrap (Bad Parts Produced)	Production Time (in Hours)
Tyrone	1	256	243	13	2.5
Tyrone	2	402	355	57	3
Rita	3	555	503	52	4
Jack	4	103	100	3	1.5
Jack	5	106	102	4	1.5
Jack	6	204	196	8	2
Gloria	7	456	430	26	4
Gloria	8	223	199	24	2
Sam	9	506	489	17	4.5
Sam	10	201	186	15	1

CHECK YOUR UNDERSTANDING

Answer each question based on the table above.

1. Which employee produced the most good parts?

    a. Tyrone   d. Gloria
    b. Rita   e. Sam
    c. Jack

2. Which employee put in the most production time?

    a. Tyrone   d. Gloria
    b. Rita   e. Sam
    c. Jack

3. Which laser was used to work the greatest number of parts in a single batch?

    a. 1   d. 7
    b. 3   e. 9
    c. 5

4. A *ratio* is a special way to compare numbers. Ratios can be used to compare the number of good parts produced to the number of bad parts (scrap) produced. To write a ratio, write the first number being compared. Then write either a slash (/) or the word *to*. The last part of the ratio is the second number being compared.

    For each laser listed on page 22, write a ratio showing the number of good parts produced to the number of scrap parts. The first one has been done for you as an example.

Laser #	Good Parts to Scrap	Good Parts/Scrap
1	243 to 13	243/13
3		
5		
7		
9		

Lasers 2 and 3 were used to process the same type of part. Examine the results from each batch. Which employee had better quality control in his or her batch?

Explain your answer.

Check your answers on page 114.

ON THE JOB

Stefania is a laser operator. She begins her shift by putting on all the necessary safety equipment. Then, she picks up a kit. Her kit includes the raw pieces that she will process with the laser. If Stefania gets a kit with 110 pieces, she must keep track of all 110 pieces. She uses quality control forms to report on the outcome for each piece.

Stefania takes one piece from the kit and loads it into the laser cabinet. She fastens the piece into place with a clip, closes the cabinet door, and starts the laser. The laser runs on its own according to the computer's directions. Stefania then opens the cabinet door. She unlocks and unloads the part from the machine. Then, she inspects the part for chips, cracks, and other flaws.

At Laser Light, Stefania can run up to three lasers at a time. She works as quickly as she can without making mistakes. She enjoys trying to beat her own records. Very often, she does.

TALK ABOUT IT
1. Discuss how visual materials, such as quality control charts, could help Stefania on her job.
2. Describe a time when a visual, such as a chart or table, helped you at home or on the job.

Businesses often change their ways of doing things. They may start using a new work method. They may put employees into teams or groups. These changes are made to help the company get better results. Companies like to see steady improvement in performance. They check for this by charting results over a period of time. In many companies, this information is presented in a graph or a chart.

Read the following example of a graph and a notice that a laser operator might see on a bulletin board.

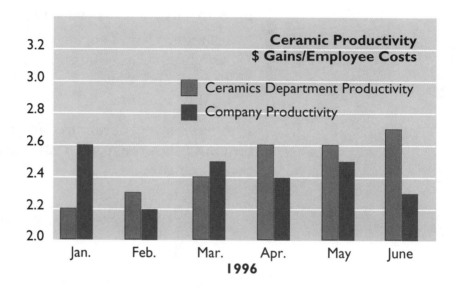

To: Ceramics Department Employees
From: Quality Control
Subject: Productivity

productivity
(proh-duhk-TIHV-uh-tee)
keeping costs and waste low,
but output high

   Each month, Laser Light measures your department's **productivity.** We compare it to the company's overall productivity. The graph shows you how your department compares with the whole company for each month so far this year. The gray bars are used to display your department's results. The purple bars are used to display Laser Light's company-wide performance. Congratulations! Your department is showing steady improvement. Please keep up the good work.

   If you have any questions about the information in this month's chart or the **procedure** we are using, please contact your supervisor.

procedure (proh-SEE-jer)
course of action

Answer each question based on the graph on page 23.

1. In which months was the ceramics department's productivity *below* Laser Light's overall rate?

    a. January and February
    b. January and March
    c. January and April
    d. January and May

2. In which months did the ceramics department's productivity rate stay the same?

    a. January and February
    b. February and March
    c. April and May
    d. May and June

3. A *trend* is a general direction or pattern. Which choice best describes the trend of the ceramics department's performance?

    a. flat; no change
    b. steady decline
    c. down and up
    d. steady improvement

4. Which of the following best describes the trend of the whole company's performance?

    a. flat; no change
    b. steady decline
    c. down and up
    d. steady improvement

5. Laser Light employees can earn raises every six months. They earn a raise if their department's productivity performance is greater than the company's performance for at least four of the six months. Did the ceramics department employees earn a raise in the January-June period?

Explain your answer.

Check your answers on page 114.

◆ LESSON WRAP-UP

In this lesson, you learned how to read information presented in several visual formats. You saw how different visual formats can be used to show the same information. You saw how tables and bar graphs can often help readers understand information very quickly.

Tables use columns and rows to organize information. Columns show information vertically. Rows show information horizontally. Specific pieces of information are within cells in the table.

Bar graphs are also good ways of showing information visually. Bar graphs can place information in a side-to-side direction. They can also place information in an up-and-down direction.

Visuals are effective tools for showing general directions, or trends. Workers can use visuals to see if productivity is increasing or decreasing. Employers can easily compare the information in different visuals.

Think about a way in which you might use a visual to show important information. Could you use a table to track how much money you spend at the grocery store each week? Could you also show this information in a bar graph?

Fill in the table at left with your *estimated* weekly grocery costs for each of the four weeks.

Now, create your own bar graph to show this information.

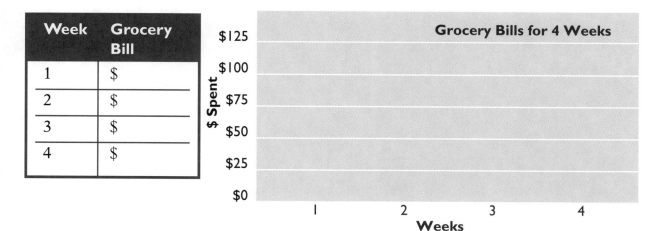

Week	Grocery Bill
1	$
2	$
3	$
4	$

Check your answers on page 114.

◆ UNIT ONE REVIEW

1. Imagine that you are at work. You pass a co-worker in the hall. She tells you that she saw a memo on everyone's desk but didn't have time to read it. She asks if you read yours. You say, "Yes." You have only a minute to explain the memo to her. What reading skill are you likely to use to explain what it said? How would you use this skill?

2. Now imagine that this co-worker comes to your work area a little later. She has more time to spend. She wants to talk about some of the specific information in the memo. What reading skill are you likely to use now? Explain how you would have used this skill.

3. Have you ever had to read a chart or diagram on the job? Have you used visuals to do things at home? Write about a time when you had to read something in a visual format.

Check your answers on page 114.

Unit Two

Jobs in Computers and Electronics

Occupations in computers and electronics are among the hottest jobs today. Packaging specialists pack and ship computers and electronic equipment. Surface mounting technicians build and test electronic parts. Desktop publishers use computers to design and produce printed materials.

In this unit, you will learn about some materials that workers in computers and electronics read. These materials include reports, notices, and schedules.

This unit teaches the following reading skills:

◆ making inferences
◆ drawing conclusions
◆ distinguishing fact from opinion

You will learn how workers in computers and electronics use these reading skills in their work.

Lesson 4

Shipping Technology Products

▼▼▼▼▼▼▼▼▼▼▼
Words to Know

packing list

product ID number

purchase order

track

Some workers in technology build new products. Once the products are built, they are sent to customers. Packaging specialists (SPEHSH-uh-luhsts) send the finished products to customers.

Packaging specialists work in shipping (SHIHP-ihng), or the sending of products. They make sure that the shipment matches the customer's order. The *number* of pieces must be right. The *types* of pieces must be right. And the materials must be sent to the right *location*.

Like other workers, packaging specialists must be good at **making inferences** (IHN-fuhr-uhns-ehz). Supervisors and co-workers may not say things directly. They expect workers to use what they already know to make correct inferences.

Packaging specialists must be able to figure out many things on their own. They may use different pieces of information to decide the best way to ship an order. If they hear a supervisor talk about an important delivery date, they should infer that the order must be there on time.

Job Focus

Packaging specialists work in company shipping departments. They are responsible for sending customers the products that they order.

Packaging specialists must pick the correct items for each order. They must pack the pieces properly so that they don't get broken. They must make sure that the order gets sent to the right customer and the correct address.

More and more technology products are sold every day. This means that more products are shipped. Jobs for packaging specialists will grow as long as technology companies continue to sell more products.

Making Inferences: How It Works

Making inferences is almost like being a detective. Detectives look for clues. So do readers who make inferences. Readers gather clues and facts. They put the clues and facts together to make an inference. This inference is something that is not stated directly in the reading.

Sometimes, you may not have all the clues you need from what you read. You will need to put information that you already know together with new information that you get from reading.

Below is an example of a notice that a packaging specialist might read on the job. As you read the notice, gather facts and clues about the company.

Notice to Shipping Department Workers

We have shipped many new orders this past month. These orders have been larger than usual. The orders are from new customers. We expect these customers to give us much more business over the next 12 months.

Our sales team expects to get even more new customers within the next two months. Rocky says he has at least three new companies who will use our products. Mel reports that she has four new stores coming on board this month. We also expect our old customers to increase their business.

The shipping department needs to keep up with the business. We need more help. If you know people interested in becoming packaging specialists, please have them call our human resources manager for an interview.

What are some inferences that you might make about the company?

_____

_____

You can infer that the *company business is growing,* or *the company is doing well.* Clues such as *We have shipped many new orders this past month* and *The orders are from new customers* help you make this inference.

To solve problems, packaging specialists must use information that they already know and also gather new clues. They often make inferences to help fill orders correctly. They use the inferences to make good decisions.

Below is an example of a shipping schedule. Packaging specialists follow this type of schedule to get orders out on time. Read the shipping schedule. Then, answer the questions that follow.

Shipping Schedule

Ship new customer orders over $50,000 first. Call new customers to let them know when their shipment will arrive. Call to make sure that the order arrived in good shape and on time.

All other orders should be shipped as soon as possible. Ship the largest order first. Ship all remaining ones in order from largest to smallest. Include a *Did Your Order Arrive as Expected?* card.

Combine orders going to the same customer at the same location. Customers may get better shipping prices. You can combine later orders with earlier orders. Use reliable shippers that give reduced prices on shipping.

Amount of Order ($)	New/Old Customer	Customer	Shipping Address	Shipping Week
$75,000	new	Ableset	112 Oak Ave. Highland Park, IL 60035	5/23
$40,000	new	Ableset	112 Oak Ave. Highland Park, IL 60035	5/23
$125,000	old	Bigmore	746 Humphrey Blvd. Redwood City, CA 94062	5/23
$157,000	old	Complt	56789 Industrial Ave. Huntsville, AL 35801	5/23
$25,000	new	DeGear	3406 Forgan Road Warehouse #7 Littleton, CO 80122	5/23
$15,000	new	DeGear	3995 Bryant Drive Warehouse #12 Chicago, IL 60601	5/23
$90,000	old	Eagen Electronics	Warehouse #7 3406 Forest Blvd. Louisville, KY 40211	5/23

Write the clues for each inference stated below. Base your answers on the shipping schedule on page 30.

Example: Inference: Big orders from new customers are the most important.

Clue 1: New customer orders over $50,000 get shipped first.

Clue 2: Ship all other orders from largest to smallest as soon as possible.

1. Inference: Combining orders is important.

Clue 1:

Clue 2:

2. Inference: The Ableset order will be shipped first.

Clue 1:

Clue 2:

3. Inference: Save the customer shipping costs.

Clue 1:

Clue 2:

4. Inference: The company is trying to please the new customers.

Clue 1:

Clue 2:

Answer each question in the space provided.

5. Whose orders will be shipped last?

What information did you use to make this inference?

6. Which customer(s) will have orders combined?

What information did you use to make this inference?

Check your answers on page 114.

Michael is a packaging specialist. He has worked at Pick Electronics for eight months. Before coming to Pick, he had been going to school part time and working part time. When his wife became ill, he needed to work full time.

Michael works the second shift at Pick. He packs and ships customers' orders. He is good at picking the right products off the shelves in the stock room. He counts and packs the correct amount of each product.

Michael packs each item in special materials. Not all materials are packed the same way. Michael must know the type of item that he is packing. He must be able to infer which types of packing materials to use and the best way to pack the material.

Once Michael has packed an order, he must restock, or refill, the shelves. If the same item is still available, he replaces it. If a new item needs to be stocked, Michael must make a new label. He must also enter this information into a computer.

Michael checks each product to make sure that the label is correct. He puts labels onto products before he packs them. He also makes the shipping labels for each box in a shipment. Michael must have good reading, writing, and speaking skills. He needs to explain any shipping mistakes to his supervisor. He must do this even if he didn't make the mistake.

Michael enjoys working at Pick. He likes his job, but he wants to get a promotion. He wants to do new tasks. He would like more responsibility and better pay. Michael knows that he needs to go back to school to get ahead.

TALK ABOUT IT

1. What inference did Michael make about furthering his career?

2. Discuss an inference you made today. What clues did you base the inference on?

Packaging specialists are happy when customers get their orders on time. They're happy when customers get everything they ordered. They're happy when they don't get complaints about orders from their supervisor.

Sometimes, mistakes happen. Then, the packaging specialist must find out what went wrong. The specialist must check that all the items were included in the shipment. The specialist must check that the items were packed correctly. The specialist must check that the cartons had the correct shipping labels. There are many more things that the packaging specialist must look at.

Below is an example of a report that a packaging specialist might write for an order that had a mistake. Read the report. Then, answer the questions that follow.

Problem Order Report_____

Date of Report: 5/30
Packaging Specialist: Michael Evans
Customer: CompIt
Customer Address: 56789 Industrial Avenue
                              Huntsville, AL 35801
Customer Phone Number: (602) 555-2432
Date of Customer Complaint: 5/30

How was the complaint made?

❑ by phone   ❑ in writing   ☑ other

If other, please explain: The complaint was made to the sales representative, Nel Connoly. Nel called me from the customer's office. She said her customer was upset. He needed the parts to finish an important order.

To whom was the complaint made? First to Nel, then to me.
Describe the complaint in detail. CompIt complained that they did not receive all the items they ordered. They checked their **purchase order**. CompIt ordered 520 connectors. The **product ID number** is C4267.

CompIt says that they only received 500 connectors. Our **packing list** shows we shipped 520 connectors #C4267. CompIt did not receive all the connectors they ordered.

What action will be taken? I will check with the shipper to **track** the problem. I will find out what went wrong with this order. I will give Nel a deadline by which CompIt will get the missing connectors.

---

purchase order
(PER-chuhs AWR-der) list of items ordered by a customer

product ID number
(PRAHD-ukt) tracking number

Packing list record of items in a shipment

track follow each step in a process

Answer each question based on the report on page 33. A question may have more than one correct answer.

1. What did Michael infer that he must do to begin solving the problem?

    a. call Nel Connoly
    b. call the customer at CompIt
    c. look up the shipper's address
    d. check with the shipper to track the problem

2. What clues gave Michael the idea that his problem needed to be corrected quickly?

    a. The sales representative called from the customer's office.
    b. The sales representative was calm.
    c. The customer was upset.
    d. The shipper made a mistake.

3. What clues did Michael use to infer that the shipper may have made a mistake?

    a. He wrote up a problem order report for the connectors.
    b. He checked the packing list.
    c. CompIt's checked their purchase order.
    d. He asked Nel to count the number of connectors.

Answer each question in the space provided.

4. What did Nel decide to do while at the customer's office?

5. If you were an employer, would you like to have Michael as a packaging specialist? Why?

Check your answers on page 115.

◆ LESSON WRAP-UP

In this lesson, you learned how to make inferences. An *inference* is information that is not stated directly. You make inferences when you read. You also make inferences at work.

When you make an inference, you act like a detective. Detectives gather clues. They add these clues to information or facts that they already know. Together, these pieces of information help them figure out what they need to know.

On the job, workers often have to make inferences. They use clues that they get in writing or in talking with someone else. They may combine these clues with the job facts that they already know. Making good inferences helps them make good decisions.

Answer each question in the space provided.

1. It's the time of the year at work when raises are given out. Your supervisor shows you a memo that she is sending to her boss. The memo lists all the extra tasks you did over the past year. It also explains how productive you were. What inference would you make from this memo?

2. Think about something you read recently. What was it? A magazine article? A memo from your supervisor? What inferences did you make while you were reading? Write a paragraph that tells about some of the inferences. You may need to reread the item before you write your paragraph.

Check your answers on page 115.

Building Electronics

▼▼▼▼▼▼▼▼▼▼▼
Words to Know

bills of materials (BOMs)

bonus

capacitor

connector

flexible schedules

gain-sharing

logic chip

resistor

Many people have home appliances. They may have coffee pots that start brewing at a specific time. They may have irons that automatically turn off. These home appliances are a type of electronics.

Many types of electronics have microchips (my-KROH-chihps). Microchips—called chips for short—are very small, thin squares or rectangles with electrical parts. They allow us to program our coffee pots. They also turn off our irons automatically. These microchips are attached to a special board called a printed circuit (SER-kiht) board.

Workers who build electronic equipment need to test the electronic parts. The workers **draw conclusions** based on the information from the test and what they already know. This means that they make decisions or judgments. For example, they decide which circuit boards will work well and which will not.

Good readers also draw conclusions based on the information they read. In this lesson, you will draw conclusions from material that technology workers read.

Job Focus

Surface mounting technicians are called SMTs, for short. SMTs work for companies that make microchips and printed circuit boards. These workers build the different parts that are used in home appliances, computers, and other types of electronic equipment. SMTs pick, place, and fasten microchips onto printed circuit boards. Then, SMTs test the circuit boards.

More and more companies use microchips in the equipment they produce. SMTs are needed to build and test the circuit boards. SMTs must have good control of their hands and fingers. They also must have good observation and problem-solving skills.

Drawing Conclusions: How It Works

Drawing conclusions is like putting together pieces of a puzzle. First, you look at the facts and details given in a reading. Then, you think about what you already know from your observations—what you see or hear around you. When you combine the information you read with the information you know, you can draw conclusions.

Surface mounting technicians need to draw conclusions when they build printed circuit boards. To build a circuit board, SMTs read **bills of materials (BOMs)**. BOMs list all the pieces that need to be fastened to the circuit board. Below is the BOM for a MaxIt Memory Card. Read the BOM.

bills of materials (muh-TEER-ee-uhlz) **(BOMs)** lists of all the parts needed for a specific job

capacitor (kuh-PAS-ih-tuhr) an electrical part that stores a charge for a short time

connector (kuh-NEHK-tuhr) a part that joins other parts

resistor (rih-ZIHS-tuhr) an electrical part that opposes the passing of an electric current

logic (LAHJ-ihk) **chip** an electrical part that uses math to find a result

Bill of Materials

Item Description: MaxIt Memory Card

Sequence	Description	Quantity	Reference Number
1	microchip	8	U1-U8
2	**capacitor**	8	Z1-Z8
3	**connector**	1	K1
4	**resistor**	6	R11-R12, R14-R17
5	resistor	5	P(3,5,8,10), R13
6	**logic chip**	2	U17, U18
7	printed circuit board	1	

An SMT has a work order to make 100 MaxIt Memory Cards. The SMT finds that there are 190 logic chips and 1000 microchips in the supply area. Based on the BOM, how many microchips and logic chips are needed to make one MaxIt Memory Card?

The SMT reads that *8 microchips and 2 logic chips are needed to make one card.* What conclusion can the SMT draw about the work order to make 100 MaxIt Memory Cards?

The SMT concludes that 10 more logic chips are needed to fill the work order. In order to make 100 memory cards the SMT needs 800 microchips and 200 logic chips.

SMTs run machines that place microchips and other pieces onto printed circuit boards. They use BOMs to know what pieces are needed to build a specific item. The BOM also tells them the order of the pieces so that they can be put on the circuit board correctly.

Each piece must go into its correct spot on the circuit board. The piece must also face in the right direction. The SMT reads a diagram to see where each piece should go. The diagram also shows which direction the piece should face.

Below is an SMT's review of a sample printed circuit board. Study the review and look at the diagrams. Then, answer the questions that follow.

M E M O

To: Margaret Hughes
From: Henry Kamin

I ran a board through the machine to see if it would work correctly. I've included a diagram of the sample board.

I examined the sample printed circuit board. Then, I compared it to the correct diagram that I had been given with the BOM. I have found three types of problems:

1. Some pieces were *not put in the correct spot* on the board.
2. All pieces *did not face in the right direction* on the board.
3. The board *did not have the correct number* of each kind of piece.

I concluded that the board wouldn't work correctly. I will need to run additional samples after correcting the problem.

Sample Printed Circuit Board   **Corrected Printed Circuit Board**

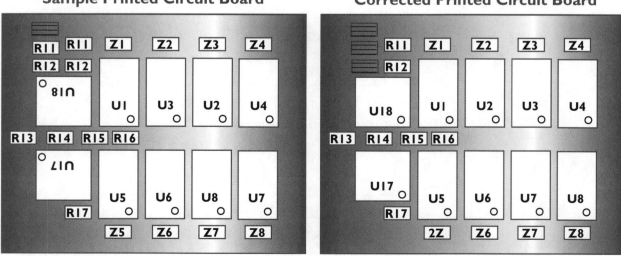

LESSON 5 ◆ BUILDING ELECTRONICS

Choose the best answer to each question that follows. Base your answers on the review and diagrams on page 38.

1. Henry's review showed that there was a problem with the logic chips U17 and U18. Look at the two boards. What problem did he catch?

    a. The logic chips were not in the correct spot.
    b. The logic chips did not face in the right direction.
    c. The board did not have the correct number of logic chips.

2. The problem with U17 and U18 is

To fix the problem, Henry should

3. Henry's review showed there were some problems with the microchips U1–U8. Look at the two boards. Which of the following was the problem?

    a. Some of these pieces were not put in the correct spots.
    b. All these pieces did not face in the right direction.
    c. The board did not have the correct number of microchips.

4. The problems with U1–U8 were

To fix the problem, Henry should

5. Henry's review showed there was a problem with the resistors R11–17. Look at the two boards. What problem did he find?

    a. Some of these pieces are missing from the board.
    b. Some of these pieces did not face in the right direction.
    c. The board did not hold the correct number of resistors.

6. The problem with the resistors is

To fix the problem, Henry should

Check your answers on page 115.

Janice is a surface mounting technician. She works for a company that builds computer parts. Janice graduated from high school two years ago. She has always liked working with her hands. She also has a good eye for detail.

Janice began working at her company as a packaging specialist. Her supervisor valued Janice's work. She saw that Janice wanted something more. Her supervisor told her to apply for the SMT Level 1 job. Janice was proud when she got it.

As an SMT, Janice runs the machines that place all the pieces on the printed circuit boards. She reads bills of materials, assembly drawings, and work orders.

Janice's eye for detail is very important on the job. Janice puts each board that she builds in the tester. She also looks over each board before it goes into the tester. She uses her observation skills to make sure that each piece is placed correctly.

Janice picks all the pieces that go onto a board. She places these pieces into bins along the assembly line. When the boards are done, Janice puts them into foam separators and brings them to her workbench. There, she tests the boards to see if they work. She tests each board that she builds.

Janice's company prides itself on producing quality parts. Janice is proud that her work reflects the company's reputation for quality.

TALK ABOUT IT

1. Why do you think that Janice is happy in her new job? Give details to support your conclusion.

2. Have you ever been proud of a job that you have done? The job may have been at home, at work, or at school. Describe the job.

SMTs are sometimes asked to work on a company advisory (ad-VY-zuh-ree) committee. An *advisory committee* is a group of people who give their ideas on how to make things better. They may also give their ideas about which things are working well and should not be changed.

The people who work on an advisory committee must be able to draw conclusions. To do so, they must gather facts, make observations, and give suggestions. For example, committee members may be asked to give their ideas about how to keep good, experienced employees.

Read the committee's report. Then, answer the questions that follow.

gain-sharing
(GAYN-shair-ihng) sharing of company profits with employees

bonus (BOH-nuhs) extra money added to one's regular salary or check

flexible schedules
(FLEHK-suh-buhl SKEH-joolz) work hours for each employee that can vary from set work times

---

Advisory Committee Report

Our committee looked at what we can do to keep good, experienced employees. We made our own observations. We also interviewed other employees to get their ideas. We have come to the following conclusions:

1. Our company has a good and fair bonus program. It is called **gain-sharing**. Each employee and each department works hard to improve its performance. When the whole company does better, all workers benefit by getting a **bonus**. The workers are very happy with this program.

2. The company has many benefits. It has good health insurance and good dental insurance. It lets employees work on **flexible schedules**, as long as the work gets done. Flexible schedules make it easy for employees to take outside work-related classes. Employees pay for these classes themselves. They feel that they shouldn't have to pay for the classes because the learning directly benefits the company.

3. Workers attend many department and work team meetings. The meetings keep them from meeting their schedules. However, employees *do* want the chance to communicate with other employees. Almost every employee believes chances for communication need to be improved.

DRAWING CONCLUSIONS

The following conclusions are drawn from each paragraph in the committee's report on page 41. Decide if each conclusion is correct or incorrect, and check off your choice. Answer the questions that follow each conclusion.

Example: Employees like the company's bonus program. It should not be changed.

☑ correct conclusion   ❑ incorrect conclusion

Which facts and observations from the report support your answer?

The workers said they are very happy with the program. All workers can share in the company's performance gains. The employees feel this is very fair and good.

1. The company has good benefits. There should be no changes made to the benefits program.

❑ correct conclusion   ❑ incorrect conclusion

Which facts and observations from the report support your answer?

2. Communication is a big problem at the company. The company must fix its communication problems.

❑ correct conclusion   ❑ incorrect conclusion

Which facts and observations from the report support your answer?

3. Describe some ideas that might fix the company's communication problems.

Check your answers on page 115.

◆ LESSON WRAP-UP

Drawing a conclusion is like putting together pieces of a puzzle. The pieces are made up of details, facts, and what you already know, including your observations. You combine this information to make a judgment or a decision.

Drawing conclusions is an important skill on the job. You may use what you see or hear. You may use something you read. The key to drawing conclusions that make sense is to choose the right facts and put them together with what you already know.

1. Has anyone ever warned you not to "jump to the wrong conclusion"? What do you think this means?

I think that "jumping to the wrong conclusion" means

2. Picture this. Your co-worker, Richard, arrives at work 15 minutes late for the third day in a row. He has a history of showing up late. What conclusion do you draw about Richard?

I think that Richard

3. You look at Richard more closely. This time you see that he is limping. His pants are torn. He is holding an ice pack to his head. Now, what conclusion do you draw?

I think that Richard

4. Describe a time when you jumped to a wrong conclusion. Explain why it was wrong.

One time

Check your answers on page 116.

Desktop Publishing

▼▼▼▼▼▼▼▼▼▼▼▼▼

Words to Know

ergonomics

font

graphics

keyboard

logo

monitor

readability

Businesses use computers to do many tasks. Some use computers to pay employees. Others use computers to keep track of their products. Many businesses also use computers to create the company's printed materials. This is called *desktop publishing.*

Desktop publishers, or DTPs, work on company newsletters, reports, presentations, and other print and visual pieces. DPTs make these printed materials attractive and easy to read. They use special computer programs to produce the materials.

DTPs make many choices when they produce a printed piece. They choose the colors, the kinds of type, and the visuals. These choices affect how the finished piece will look.

DTPs use facts and opinions to make choices. A *fact* is something that can be proven. An *opinion* is a personal judgment or belief. Different people may have different opinions about the same thing. Like all workers, DTPs must be good at **distinguishing fact from opinion.**

Job Focus

Desktop publishers must know how to use many computer software programs. They may read magazines, manuals, and books to learn new computer skills. DTPs also read these materials to learn new design skills. *Design* refers to the "look" of a piece.

Some companies need full-time DTPs. Many full-time DTPs work for printers or for advertising firms. Other full-time DTPs work in company marketing departments.

Some companies use part-time DTP services to design and produce a specific piece. A DTP may work for a DTP service or may be self-employed.

Distinguishing Fact from Opinion: How It Works

A *fact* is information or a statement that can be proven true. For example, a DTP has just finished making the company's newsletter. One statement about the newsletter might be: "The newsletter is four pages long." This would be a fact. The pages could be counted, and the statement could be proven true.

When you talk about what you believe or when you judge something, you are giving your *opinion*. For example, "This newsletter is beautiful" is an opinion. You may agree that it is beautiful, but the statement is an opinion, not a fact. Someone else may have a different idea about the newsletter.

To **distinguish fact from opinion,** ask yourself, "Can this be proven to be true?" If it can, it's likely to be a fact. You can also ask yourself, "Is there more than one way to think about this?" If there is, it's likely to be an opinion.

Read the supervisor's note with comments about a DTP's work.

Dear James:

    I really like this piece. You used all the company colors in the piece. You made our blue brighter. We should use this new blue from now on. We will need to get an okay from our president to make this permanent change.

    You also moved our company **logo** to the middle of the page. This looks good. You are using good design skills. Good job!

                  Thanks, Pat

---

logo (LOH-goh) symbol that a company or group uses to identify itself

What are two facts from the note?

1. _____

2. _____

What are two opinions from the note?

1. _____

2. _____

You may have found these two facts: *(1) James used all the company colors,* and *(2) he moved the logo to the middle of the page.* You may have found these two opinions: *(1) Pat likes the brighter blue,* and *(2) she likes James's work.*

DTPs combine text and visuals to make their work attractive. They must also make their work easy to read. They want their audience to be able to understand what they are looking at and reading. These are the main goals of a DTP.

To meet their goals, DTPs make choices. They choose where to place the words and the visuals. They choose the type style for the text.

DTPs make dozens of choices to produce just one piece. Many choices are based on the DTP's own opinions. Other choices are made based on facts that a DTP knows.

Below is a magazine article that a DTP might read. The article talks about good design rules. Read the article. Then, answer the questions that follow.

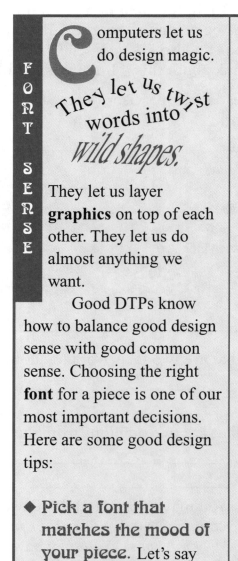

FONT SENSE

Computers let us do design magic. They let us twist words into *wild shapes.* They let us layer **graphics** on top of each other. They let us do almost anything we want.

Good DTPs know how to balance good design sense with good common sense. Choosing the right **font** for a piece is one of our most important decisions. Here are some good design tips:

◆ **Pick a font that matches the mood of your piece.** Let's say you are making a poster for a carnival. Pick a font that is fun and happy. Don't use a font that is too serious, gloomy, or sad.

◆ **Limit the number of fonts you use in a single piece.** One of the biggest mistakes is using too many fonts in a single piece. Research has shown that this confuses the readers.

◆ **Make your type comfortable to read.** The type *style* you choose is important. So is the type *size*. If type is too large or too small, it will make things hard for your readers. Think about how your choices will affect the **readability.**

graphics (GRAF-ihks) pictures or other visuals

font (fahnt) style for printing letters, numbers, and symbols

readability (reed-uh-BIHL-ih-tee) ease or difficulty of reading material

Read each statement below. If it is a fact, circle *fact*. If it is an opinion, circle *opinion*. Base your answers on the article on page 46.

fact   **opinion**   **1.** Computers can twist words into wild shapes.

fact   **opinion**   **2.** Computers let DTPs layer graphics on top of each other.

fact   **opinion**   **3.** Choosing the right font is a DTP's most important decision.

fact   **opinion**   **4.** The font should match the mood of the piece.

fact   **opinion**   **5.** The size of the type will affect the piece's readability.

Read the following statements. Decide if each is a fact or an opinion. Explain your answer in the space provided.

Example: Using too many fonts in a single piece is a big mistake.
   a. *opinion*
   b. *This is how the writer of the article feels. How many is "too many?" The answer could vary from person to person. Also, some materials may use more fonts than others.*

6. Serious fonts are poor choices for posters.
   a.

   b.

7. Computers can do design magic.
   a.

   b.

8. Using too many fonts in a single piece confuses readers.
   a.

   b.

Check your answers on page 116.

Betty loves computers. She spends most of her work day using one. Betty works as a desktop publisher.

Betty began taking adult education classes at her local high school when she was eighteen. She worked hard and earned her GED (General Equivalency Diploma).

Betty learned how to use a computer in her adult education classes. Whenever she had free time, she would go to the computer lab. She first learned the word-processing (PRAHS-ehs-ihng) programs. She used these programs to write and type stories and letters. Then, she taught herself how to use many other software programs. These programs let her make the design of her pieces more interesting.

After getting her GED, Betty got a job with a local charity. Betty uses the computer to produce the charity's printed materials, including its newsletter. Betty also uses the computer to make posters, fliers (hand-outs), and reports.

Betty uses a variety of software programs to make these materials. She often uses a word-processing program for simple reports. For more difficult items, Betty also uses page layout and graphics software programs. These software programs let her make materials that really grab people's attention.

Betty has strong computer skills. She knows how to use the computer and software to put text and graphics on the page. Betty also reads lots of magazines on design and desktop publishing. The magazines give her new ideas and keep her skills up to date.

Next fall, Betty plans to return to school. She will be taking computer design classes at the local college.

TALK ABOUT IT

1. DTPs need good communication skills. Discuss times that Betty might have to use these skills.

2. Describe a time when you had to deal with two or more opinions. What did you do?

DTPs work at their computers most of the day. They spend most of the day sitting in the same position. They tend to repeat the way they move their wrists, fingers, and hands. This can cause health problems.

Following is a page from a booklet for DTPs. It gives information on **ergonomics**—the research and practice of helping people work safely in a healthy workplace. Read the page. Then, answer the questions that follow.

ergonomics
(air-goh-NAHM-ihks) field of safe and healthy work practices

Ergonomics Guide
for Desktop Publishers

The equipment you use at your workstation must be adjusted specifically for your body. The diagram below shows the major adjustments you should make. Be sure your equipment is set to these guidelines.

Body Check
1. Eyes—your eyes should be looking down slightly at the **monitor**. This protects your eyes from glare and stress.
2. Wrists—your wrists should be flat with the top of your hand at the middle of the **keyboard**.
3. Back—your back should be in a comfortable "S" shape. Do not slouch or lean to the side.

Equipment Check
A. Monitor—the monitor should be an arm's length (about 18–24") from your eyes.
B. Keyboard—the keyboard should be at the same height as your elbows.
C. Chair—the back rest of the chair should support your lower back. The seat height should make your knees form a 90° angle.

monitor
(MAHN-ih-ter) screen on the computer, where you see your work

keyboard (KEE-bawrd) part on the computer with keys for typing

CHECK YOUR UNDERSTANDING

Answer each question based on the booklet on page 49.

1. Which of the following statements is an opinion?
   a. People can injure themselves working with computers.
   b. People should work in a safe workplace.
   c. Ergonomics can make workplaces safe.

2. Which of the following statements is a fact?
   a. DTPs have the best job in computing.
   b. DTPs do not need to think about their health.
   c. DTPs can adjust their equipment to improve their health.

Read each statement below. If it is a fact, circle **fact.** If it is an opinion, circle **opinion.**

fact   opinion   3. All computer operations need to be ergonomically improved.

fact   opinion   4. If your wrist is higher than your knuckles when you work at a computer, you may get a hand problem.

fact   opinion   5. Desktop publishers produce too many booklets.

fact   opinion   6. Booklets provide helpful information.

fact   opinion   7. Slouching or leaning while working on the computer can hurt your back.

Read the note below. Pick a fact and an opinion that Billy included. Write them in the space provided.

8. Dear Holly:

I just got this great booklet. It shows lots of information about how we can work safely at the computer. I am giving you a copy of this ergonomics page. I suggest that you read it. We will have a safer workplace if we follow these guidelines.   Billy

fact:

opinion:

Check your answers on page 116.

LESSON 6 ◆ DESKTOP PUBLISHING

◆ LESSON WRAP-UP

In this lesson, you learned about facts and opinions. You learned that:

- a **fact** is a statement that can be proven true.
- an **opinion** is somebody's belief or judgment about something.

It can be tricky to decide if a statement is a fact or an opinion. You must read the statement carefully. There are two good tests to use to tell a fact from an opinion.

Test 1: Ask yourself, "Can this be proven true?" If it can, it's likely to be a fact.

Test 2: Ask yourself, "Is there more than one way to think about this?" If there is, it's likely to be an opinion.

Distinguishing fact from opinion is an important skill to have. You may need to make a suggestion on the job. If you do, you will be stating your opinion. You will probably need to support your opinion with facts.

Picture this. You are a desktop publisher. You sit between two other DTPs. The two other DTPs don't get along. They mumble things about each other. They give each other cold stares. Their problem is beginning to affect your work. You want to ask your supervisor to move your workstation.

In the space provided, write your supervisor a request to be moved. Your request should be stated as an opinion. Then, give your supervisor some facts to support your request.

Dear Supervisor:

Sincerely,

(Your Name)

Check your answers on page 116.

◆ UNIT TWO REVIEW

1. A co-worker tells you that your work area should be clean at the end of the day. Is he telling you a fact or his opinion? Explain how you would tell.

2. Imagine that the co-worker's comment above was an opinion. What kind of conclusion might you draw about him? What might you infer about his behavior?

3. Have you ever had to tell the difference between a fact and an opinion? Talk about a time at work when you had to use this skill.

Check your answers on page 117.

Unit Three

· Communications · Technology Occupations

Working with communications (kah-myoo–ni-KAY-shuhnz) technologies can be exciting and fun. Mobile communications technicians set up and care for the phones that keep people in touch while they are on the move. Photo lab technicians develop pictures for individuals and businesses. Satellite dish installers set up and maintain satellite dishes that bring dozens of TV stations into people's homes.

In this unit, you will work with the types of information sheets, phone logs, and written directions used by workers in communications technology.

This unit teaches the following reading skills:

◆ comparing and contrasting
◆ identifying cause and effect
◆ following directions

You will learn how workers in communications technologies use those reading skills in their work.

Working with Mobile Phones

▼▼▼▼▼▼▼▼▼▼▼

Words to Know

adapter

cellular phones

microphone

models

portable

talk time

watts

Have you ever seen drivers talking on a telephone in their cars? Have you seen someone walking down the street and talking on the telephone? Both of these situations are part of a growing trend called mobile communications.

Technology has made it possible for people to talk on the phone while in their cars. With new technology, people carry phones with them as they walk down the street. Using wireless phones, people communicate while they're in motion. The popularity of wireless phones has led to jobs for mobile communications technicians. Phones that are mobile (MO-buhl) are movable or portable.

Think about your phone at home. Your phone is plugged into a jack in the wall. It is called a "hard-wired" phone. Wireless and "hard-wired" phones are similar. People can talk with one another, either locally or long distance, using both types. However, there are big differences between these two types of phones. Wireless phones let you travel almost anywhere while you talk. Hard-wired phones keep you close to the receiver.

Looking at the similarities and differences between things is called **comparing and contrasting**. You compare the qualities that are the same. You contrast the qualities that are different. This skill is important for technicians who work with telephone customers.

Job Focus

Mobile communications technicians (MCTs) install, service, and fix phones and fax machines. They make sure that phones and fax machines work correctly. They also help customers learn how to use wireless phones.

Mobile communications is getting more popular. People are often on the go and need to keep in touch with others. Skilled MCTs will have jobs as long as people continue to use wireless products.

Comparing and Contrasting: How It Works

Clue Words	
Compare	**Contrast**
alike	different from
both	
in common	in contrast
the same as	but
like	less
similarly	more
corresponding	on the other hand

Comparing and contrasting is a skill that workers use on the job. When workers look at ways things are alike, they are *comparing*. When workers look at the differences among things, they are *contrasting*. Technicians use this skill when they work with different kinds of products.

Clue words are often used to show if things are being compared or contrasted. The chart at left shows some of these clue words.

MCTs use phone logs to record talks they have with customers. Read the two log entries below.

PHONE LOG

Customer: Josie Logan   **MCT:** Jeremy   **Date:** 6/26

Ms. Logan is a single woman. She is a saleswoman. She needs a phone she can use in her car for work. She also wants the phone for her own safety. She is on the road alone four days a week.

Customer: Sally Cortesi   **MCT:** Jeremy   **Date:** 6/26

Mrs. Cortesi is a working mom with two children. Mrs. Cortesi wants her children's school to be able to call her at any time. She is concerned about her children's safety. Mrs. Cortesi needs a phone she can use in her car. She also needs to be able to take the phone with her on foot.

Name two ways that these customers are similar. Use comparing clue words from the chart.

_____

_____

Did you say that *both customers are women* and that *they both want a phone for safety reasons?* These are the similarities.

Name two differences between these customers. Use contrasting clue words from the chart.

_____

_____

You may have said that *Josie Logan just needs a phone in her car, but Mrs. Cortesi needs a phone to carry with her. Ms. Logan needs a phone mainly for her work. On the other hand, Mrs. Cortesi needs a phone mainly for family reasons.* These are differences.

models (MAHD-ehlz) different styles or types of a product

cellular (SEHL-yoo-ler) **phones** telephones that operate on radio waves

MCTs must know how to use different **models** of phones. For example, there are many companies that make **cellular phones**. Some cellular phones are installed in cars. Some can be carried in a purse, briefcase, or pocket. Some cellular phones are too big for a purse, briefcase, or pocket.

Read the advertising sheet below. Then, answer the questions that follow.

We're getting more mobile every day. Many of us don't work in an office. We work out of our homes or "in the field" and aren't tied to a desk. We need phones that will let us communicate with our offices, customers, and families.

portable (PAWRT-uh-buhl) able to be carried easily

watts (wahts) units of electrical power

talk time amount of time that you can talk on a mobile phone without having to recharge the battery

Wide Open Communications has phones for every mobile need. How mobile do you need to be?

This hand-held **portable** can be carried in a pocket or purse. It weighs only 18 oz. You get .6 **watts** of power and 45 minutes of **talk time.** If you need a small, easy-to-carry phone, this one is for you.

This phone takes a small battery that must be recharged often.

Model GC1776 Flip-Fold Phone - $55

Our newest bag phone can be taken wherever you need it. The phone weighs 3 pounds; with 3 watts of power and 1 hour of talk time. If you want a phone for the car and the beach, this is it.

This phone uses a powerful battery for greater cellular range.

Model NU1200 Bag Phone - $60

microphone (MEYE-kruh-fohn) a device in a telephone that changes sound into electricity to be sent to another telephone

This model includes a hands-free **microphone.** It lets you keep both hands on the wheel while driving. The handset weighs 18 ounces. You get 3 watts of power and unlimited talk time in the car. Note: This is not a portable model.

Model GC1666 - Car-Mounted Phone - $120 (installation included)

Answer each question based on the advertising sheet on page 56.

1. Which two phones are similar in price?

    a. flip-fold phone and bag phone
    b. flip-fold phone and car-mounted phone
    c. bag phone and car-mounted phone

2. Which phone is the *least* powerful?

    a. car-mounted phone
    b. bag phone
    c. flip-fold phone

3. Which quality was not compared among the three phones?

    a. weight of the phones
    b. size of the phones
    c. amount of talk time
    d. battery power

4. What is the greatest difference between the flip-fold phone and the bag phone?

    a. The flip-fold phone rarely needs recharging.
    b. The flip-fold phone can be used by workers.
    c. The flip-fold phone weighs much less.

Read the customer descriptions below. Then, decide which phone would be best for each customer.

5. I need a phone for work. I also need a phone to carry at lunch and on my way home. I want a phone that's light and easy to carry. I need to take the phone into stores with me.

    a. flip-fold phone
    b. bag phone
    c. car-mounted phone

6. I do business with companies all across the United States. In my spare time, I golf, sail, and play tennis. I want a powerful phone that I can use when I do all these things. I don't care how big the phone is.

    a. flip-fold phone
    b. bag phone
    c. car-mounted phone

Check your answers on page 117.

Jeremy James, better known as J.J., loves to read. Each day, he reads the newspaper. He reads four magazines a week. J.J. enjoys learning on his own.

J.J. started working at Best Deal, a discount electronics store. At Best Deal, he got on-the-job training from Louis. Louis taught J.J. about electronics, circuits, and wiring. He made sure that J.J. had good mechanical skills.

Then, Louis left Best Deal to work for Wide Open Communications. He saw that Wide Open needed another MCT. Louis thought J.J. would be good for the job.

J.J. got the job. He got a lot of training on this job from Louis, too. At first, J.J. watched and assisted Louis. Soon, he was installing and repairing equipment on his own. Now, J.J. installs, repairs, and services all types of mobile communications equipment.

Installing car phones is J.J.'s favorite job. J.J. has learned it's important to show the customer where the phone will go before he begins installing it. He installs three main parts of the car phone. J.J. knows where to put the antenna and hands-free microphone. The handset can go on the console between the two front seats, on the dashboard, or on the pedestal in the cradle. It depends on the design of the car.

Once the phone is installed, J.J. programs it with its own number. He tests all the phone's features. Then, he shows the customer how all the features work.

TALK ABOUT IT

1. Use the compare clue words on page 55 as you describe what J.J.'s two jobs have in common.

2. Explain how an MCT could use comparing and contrasting skills.

P art of an MCT's job is to show customers how to use their mobile phones. When MCTs show customers how to use the phone, they compare and contrast. They also give the customer an instruction sheet. This sheet tells the customer how to use the features on the new phone. It also shows some of the differences between a mobile phone and an ordinary phone.

Read the information sheet below. Then, answer the questions that follow.

adapter (uh-DAP-ter) device for connecting parts that otherwise would not fit

---

Mobile Phone Customer Information Sheet

Congratulations on your new mobile phone. Please keep this information sheet. It is a record of how to use features that come with your new phone. Look at what your mobile phone offers.

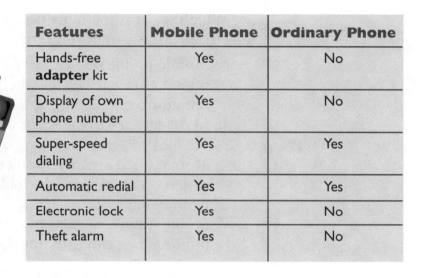

Features	Mobile Phone	Ordinary Phone
Hands-free **adapter** kit	Yes	No
Display of own phone number	Yes	No
Super-speed dialing	Yes	Yes
Automatic redial	Yes	Yes
Electronic lock	Yes	No
Theft alarm	Yes	No

Hands-free adapter kit: This kit lets you plug the mobile phone into your car.

Display of own phone number: The 7-number display lets you see your own phone number. This convenient feature will ensure that you never forget your own mobile phone number.

Super-speed dialing: Your mobile phone lets you enter up to 50 phone numbers for speedy dialing—more than any ordinary phone. You also can enter a person's name with the phone number.

Automatic redial: Your mobile phone will automatically redial any busy number you try.

Electronic lock: This helps you control the use of your phone. With an ordinary phone, you are only charged for calls you make. With a mobile phone you are charged for calls you make and for calls you receive. The lock lets you keep others from using your phone and running up costly bills.

Theft alarm: This in-car feature helps keep your phone and car safe. Program the alarm with a phone number you choose (police, home, etc.). When the phone is armed, the number will be called when your car is started. You will need to disarm the alarm before you start the car.

CHECK YOUR UNDERSTANDING

Answer each question based on the information sheet on page 59.

1. Which two features do mobile and ordinary phones share?

    a. hands-free adapter kit
    b. display of own phone number
    c. super-speed dialing
    d. automatic redial
    e. electronic lock
    f. theft alarm

2. Which set of statements best explains why the electronic lock is important for a mobile phone owner?

    a. Mobile phone owners may forget their own phone number. The lock helps them remember the number.
    b. Mobile phone owners are very busy. They can use the lock to turn the phone on and off.
    c. Mobile phone owners don't want their phones stolen. They can use the lock to keep the phone from being stolen.
    d. Mobile phone owners pay for any call on their phone. They can use the lock to control their bills.

For each question, write your answer in the space provided.

3. Contrast why the theft alarm feature is a valuable feature for a mobile phone but not for an ordinary phone.

4. Compare why super speed dialing is a valuable feature for both mobile and ordinary phones.

5. Why is automatic redial an important feature to have on a mobile phone used in the car?

Check your answers on page 117.

◆ LESSON WRAP-UP

When you compare and contrast, you find differences or similarities between two or more things. You can compare and contrast items, ideas, and qualities.

When you are comparing, you look at how items are alike. They may be alike because they are used for the same purpose. For example, you saw how people on the go use three types of mobile phones to help them keep in touch with others.

When you are contrasting, you look at differences. For example, you contrasted the qualities of three mobile phones. One could be used only in a car. Another had limited use because its battery was not strong.

You also learned about words that signal similarities and differences. Some of these clue words can help you find what is being compared. Others will help you see what is being contrasted.

Use your imagination to picture this situation. An MCT needs to install a car phone for a customer. He has installed phones in the same kind of car before. However, he's never installed this model of the phone because it's new. Answer the questions below.

I. What similarities might he look for between this installation and other ones he did in the past? (Hint: Compare the current car with a previous car.)

2. What differences might he look for between this installation and a previous one? (Hint: Contrast the new phone with older models.)

Check your answers on page 117.

Developing Film

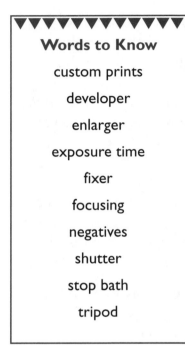

▼▼▼▼▼▼▼▼▼▼▼▼

Words to Know

custom prints

developer

enlarger

exposure time

fixer

focusing

negatives

shutter

stop bath

tripod

Do you use a camera? Do you take pictures of friends? Do you take pictures of your family? If you do, then you probably bring your film to a photo lab.

Photo lab technicians (PLTs) develop film into prints, slides, or negatives. They use timing, temperature control, and chemicals to process, print, and store the film.

Photo lab technicians know that things such as light and film speed affect how the photo will look. The PLT can make changes in timing and temperatures to improve or change the way the final photo looks.

The results the photo lab technician gets in the final photo are the *effects*. The steps the PLT takes while processing the photo are the *causes*. **Identifying cause and effect** is an important skill for PLTs.

Job Focus

Photo lab technicians (PLTs) develop film, and they print and store the photos people take. They develop film for black-and-white prints, for color prints, and for slides. When PLTs develop and print film, they use special equipment.

PLTs work in a variety of places. They can work in large film-processing labs or for printing companies. Many work in 1-hour photo stores or operate mini-lab equipment in photography stores. Jobs for PLTs can also be found in photography studios.

The job outlook for photo lab technicians is good into the next century. Jobs are expected to grow about as fast as the national average for growth in all jobs.

Identifying Cause and Effect: How It Works

A *cause* is an action that makes something happen. What happens is called the *effect*.

Photo lab technicians must be good at **identifying cause and effect**. They must read about methods for developing different kinds of film. Each step in a method produces an effect. Each step also produces a new cause.

The chart below tells the PLT the purpose of each step in developing film. Read the chart.

developer (dih-VEHL-uhp-er) chemicals used to make the image appear in film

stop bath chemicals used to stop developing the film image

fixer (FIHK-sehr) chemicals that clear off unused chemicals in film

Processing Step	Purpose
1 **Developer**	uses the chemicals in the film to make the picture (image)
2 Wash or **Stop Bath**	stops the development of the image quickly and evenly
3 **Fixer**	dissolves any undeveloped chemicals in the film; does not affect the developed image
4 Wash	removes fixer and dissolved chemicals
5 Clearing Agent	quickly removes any chemicals left on the film; shortens rinsing time from 20 to 5 minutes
6 Wetting Agent	helps film dry evenly and quickly; reduces possibility of air bubbles drying on film
7 Drying	makes image permanent on film

What is the effect of using the developer?

_____

What causes the film to dry evenly and quickly?

_____

You should have said that *the developer causes the image to appear on the film.* You also should have said that *the wetting agent causes the film to dry evenly and quickly.*

Which column of the chart would you label causes? Which column of the chart would you label effects?

_____

_____

The Processing Step column could be labeled "causes" and the Purpose column could be labeled "effect."

Some photo labs have help sheets for their customers. SPLTs look at each roll of film they develop. If they see problems, they enclose a help sheet with the customer's order.

Read the help sheet below.

TAKING BETTER PHOTOGRAPHS

Thank you for choosing **Big Finish Photo Labs** for your photography needs. Our photo tip sheets can help you take the best pictures possible.

Photo Tip Sheet #1

Problem	Reason	How to Fix It
Blurred, fuzzy pictures	**1.** Moving the camera when you snap the picture	Be sure to hold your camera very still. Use a **tripod** or a ledge, or brace your arms to steady the camera.
	2. Dirty lens	Clean your lens with lens tissue paper and cleaning solution.
	3. Poor **focusing**	Use the focusing aid in your camera correctly.
	4. Wrong **shutter** speed	If you're holding the camera in your hand, don't use shutter speeds slower than $\frac{1}{30}$ second for still shots. For shots of moving subjects, use the fastest shutter speed you can.
Red or white eye in flash pictures	Flash reflects in the subject's eyes	Turn on all the lights in the room. This will make the subject's pupils contract (get smaller). The pupil is the part of the eye pictures that reflects the flash.
Flash pictures are too light or too dark	**1.** Pictures are too light—flash is too close to the subject **2.** Pictures are too dark—flash is too far from the subject	Be sure to check the distance between your subject and the camera. You can use the focus setting on your camera to help you.
Glare spots in flash pictures	Shiny surfaces reflect the light from the flash	Take flash pictures at an angle to the shiny surface. If your subject wears glasses, have him or her turn his or her head slightly, or take off the glasses.

tripod (TRY-pahd) three-legged support used to hold a camera

focusing (FOH-kuhs-ihng) moving the lens of a camera to make the subject clearer

shutter (SHUT-ehr) part of the camera that opens and shuts to let light in or to keep it out

The list of causes below has been taken from the photo tip sheet on page 64. Use it to choose the best answer to each question.

I. The camera is too close to the subject.
II. The shutter speed is wrong.
III. The subject is wearing glasses.
IV. The subject is looking directly at the flash.

1. What causes glare spots in pictures?
a. I
b. II
c. III
d. IV

2. What produces a fuzzy effect in a photo?
a. I
b. II
c. III
d. IV

3. What should customers do to prevent the red-eye problem in their pictures?
a. Turn out all the lights.
b. Shrink the subject's pupils.
c. Take the flash off the camera.
d. Turn on all the lights.

For each question, write your answer in the space provided.

4. What effect can a subject wearing glasses produce in flash photographs?

5. Explain what a photographer can do to fix this problem.

Check your answers on page 117.

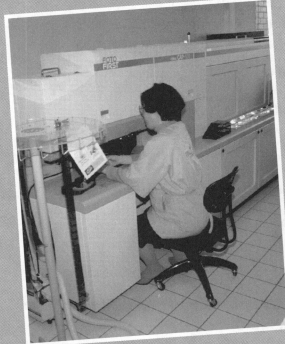

Rashida took a photography course at the local community college. In the course, she learned how to develop film and print pictures. Rashida liked it so much that she looked for a job as a photo lab technician.

Rashida found work at a 1-hour photo mini-lab. They trained her on the job. They were pleased that Rashida brought some developing experience and knowledge with her.

Rashida later moved to a job at Big Finish Photo Labs. Big Finish processes black-and-white, color print, and color slide film for discount stores. It also processes film for grocery stores, drug stores, and photography stores.

Big Finish processes pictures for professional photographers, too. They also make banners, posters, and huge prints for advertising agencies and large companies.

Rashida must work quickly and carefully. She often processes dozens of rolls of film a day. Rashida checks and reads each film cartridge before developing the film. Each film requires its own chemicals, time, and temperature settings. The settings affect how the processed film will look. The wrong settings can cause unwanted effects. She also inspects each roll for dust and air bubbles. These problems can cause damage to the customer's film, prints, or slides.

Rashida runs many machines to process and print photos. She even uses a computer to retouch and enlarge some photos. Rashida is eager to keep learning new photofinishing skills.

TALK ABOUT IT
1. Explain how Rashida could use cause-and-effect skills on the job.

2. Describe the different types of materials that you think Rashida reads on the job.

custom prints
(KUS-tuhm prihnts) photos
developed in a special way
for a customer

negatives (NEHG-uh-tihvz)
developed film

enlarger (ihn-LAHRJ-er)
machine that projects an
image through a lens to
resize the image

exposure (ihk-SPOH-zher
tym) time number of
seconds needed for light to
pass through a negative

Some PLTs work with film in large labs. They often use automatic equipment to develop film and print standard-sized pictures. Other PLTs work in specialty labs. They use a variety of materials and equipment to make **custom prints** on special paper from **negatives.**

Print development starts with controlled light. An **enlarger** controls the amount of light passing through the negative and onto special paper. This light is measured in **exposure time.**

PLTs must judge and inspect the work they do. Below is a sample of a guide for judging the quality of photos. Read the guide. Then, answer the questions that follow.

PHOTO EVALUATION GUIDE

QUALITY OF THE PHOTOGRAPH	PROBLEMS	SOURCES AND CORRECTIONS
Density—lightness or darkness of whole print	1. Too much—whole print looks too dark 2. Too little—whole print looks too light	1. Too much silver laid on the paper. This comes from too much exposure time. Adjust exposure time. 2. Too little silver laid on the paper. This comes from too short an exposure time. Adjust exposure time.
Sharpness—overall clarity of print	Print looks fuzzy	Enlarger may not have been focused at time of exposure. Or enlarger may have been shaking at time of exposure. Check test print for sharpness or make another print.
Contrast—the difference between the shadows (dark areas of the print) and highlights (light areas)	1. Too harsh—shadows are too dark and highlights are too light. Detail is lost. 2. Too flat—not enough difference between highlights and shadows. Print looks dull, cloudy, or gloomy.	1. Negative used to print photo may have too much contrast. Use lower contrast negative or paper. 2. Negative used to print photo may not have enough contrast. Try higher contrast negative or paper.
Spots—white or dark flaws on the print	1. White spots 2. Dark spots	1. White spots come from dust on the negative or paper during printing. Dust blocks light from getting to the paper. The paper color stays visible. Make sure negative and paper are dust-free before exposing to light. 2. Dark spots come from scratches or dust on the negative during its developing. This leaves a blank spot. During printing, light passes through the spot onto the paper. Scratch dark spots off print with sharp razor or bleach them off.

Answer each question based on the Photo Evaluation Guide on page 67.

1. Adjusting the exposure time affects the print's
   a. density
   b. sharpness
   c. contrast
   d. spots

2. What effect does the contrast level have on the print?
   a. Contrast makes dark spots on the print.
   b. Contrast helps the overall print look clearer.
   c. Contrast puts too much silver on the print.
   d. Contrast sets shadow areas apart from light areas.

3. Which of the sources below could have caused spots to appear on the print? (Choose *two* causes.)
   a. Scratches left blank spots on the negative.
   b. The negatives didn't have much contrast.
   c. Dust blocked light from getting to the paper.
   d. The enlarger was not focused.

4. Which of the following would **not** affect the overall sharpness of the print?
   a. the enlarger being out of focus
   b. a jiggling enlarger
   c. the sharpness of the negative
   d. scratches on the negative

For this question, write your answer in the space provided.

5. Why is it important for PLTs to keep all equipment and the darkroom clean and dust-free?

Check your answers on page 118.

◆ LESSON WRAP-UP

In this lesson, you learned how to find causes and their effects. You learned that a cause is the action that leads to a result. The result is called the effect.

Read the following example.

Eric mixed chemicals for developing film. The label on the chemical container warned Eric to wear rubber gloves. He forgot. When he splashed some chemicals on his skin, he got a painful burn. While taking care of the burn, he overexposed the film. As a result, the pictures were too light.

What was the final effect of Eric's forgetfulness? *The pictures came out too light.* What was the first cause of Eric's problem? *He forgot to put on rubber gloves.*

What else caused Eric's problem? Did you say *splashing harsh chemicals on his skin?* What else happened? Did you say *overexposing the film?*

The example above shows how one cause leads to another—and then to a final effect.

Read and answer the questions below.

1. What information on warning labels would be helpful in understanding cause and effect? Think about warning labels you've read for medicines you've taken or given to your children.

2. Why is it important to have strong cause-and-effect reading skills on the job?

Check your answers on page 118.

Installing Satellite Dishes

▼▼▼▼▼▼▼▼▼▼▼

Words to Know

alignment

equator

feedhorn

ground rod

LNB

obstructions

ordinances

plumb

Satellite (SAT-ul-yt) dishes bring TV, radio, telephone, and other services to people who live far from cities. Dishes give customers dozens of television and video channels.

In many ways, satellite technology has made the world smaller. It puts us in touch with people and events around the world. It lets us get the information and entertainment we want—when we want it.

Today more people are buying satellite dishes. More jobs are now available for people who want to work with this technology. Some of these jobs are for workers who install satellite dishes. Satellite dish installers **follow directions** on the job. These workers need to follow written directions, spoken directions from co-workers, and directions on diagrams. This is important because satellite dishes must be set up in a correct manner.

Job Focus

Satellite dish installers (SDIs) set up dish systems, also called *earth stations*, for customers. They pick the right site for the dish. They put together the dish and its parts. They point the dish in the right direction. This must be done with care, so that the dish gets signals from a satellite. SDIs also set up the cables that connect the outside dish to the inside TV monitor.

SDIs must have good math skills. They must also be able to read diagrams and charts. This helps the dish installer tune the dish system. Because SDIs sometimes need to lift heavy objects or climb difficult heights, they should be healthy and strong.

The job outlook for dish installers is good. This outlook will remain good. More individuals and businesses are asking for satellites dishes every day.

Following Directions: How It Works

Directions describe how to do something. You follow one or more steps to complete the directions. Here's one strategy that can help you understand directions.

1. Read the directions all the way through. Read all of the directions. Study all visual information—such as diagrams—that come with them.
2. State the final result. Ask yourself, "What will I accomplish by doing all the steps?"
3. Think out each step that you must follow. Read the directions again. Write the steps in your own words to make a guide.
4. Do the steps. As you complete each step, check it off on your guide.
5. Check your work. When you finish all the steps, check the final result. Is it what you planned? If not, look over your outline and the directions again.

Satellite dish installers must find the right place to set up the dish. Read the directions below.

Choosing the Right Site

When picking the right site for the dish, you must think about many things. Most important, the dish must be able to face the **equator** without any **obstructions**. Mountains, hills, buildings, trees and other objects can block satellite signals. For installations in the United States, the dish must face south.

Think about the season in which you are installing the system. If it is in winter, the trees may be bare. In summer, the leaves can interfere with the dish's reception. Even the smallest obstructions will block the signals from the satellite.

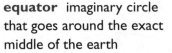

You must also check each town's rules or **ordinances** for dish locations. Some communities do not allow dishes in front of homes. Some may not allow them on rooftops.

---

equator imaginary circle that goes around the exact middle of the earth

obstructions objects that block other objects

ordinances local government rules or laws

What might happen if an SDI did not read all the directions before starting to install a dish?

_____

_____

You may have said that *the SDI may put the dish in a site that is against town rules.*

Dish installers must be able to follow directions exactly when they set up earth stations for their customers. This is very precise work.

The dish is used to collect the signals from "birds," or satellites in space.

Read the directions below for installing a dish.

ground rod part of a satellite dish system that protects it from lightning strikes and electrical problems

plumb (plum) place straight up and down and side to side

feedhorn part of a satellite dish system that collects signals that bounce off the dish

LNB part of the satellite dish system that helps process signals

Installing the Earth Station—Ground Mounting the Dish on a Pole

1. Make a secure base for the pole. Dig a hole 3–4 feet deep. Mix the concrete and pour it into the hole. Place the mounting pole in the hole. You also will need to place the **ground rod** in this base.

2. **Plumb** the pole before the concrete sets. The pole must be totally vertical or the dish will not track the signal correctly. Use a plumb rule or line.

3. Mount the **feedhorn** and **LNB** to the dish. Be sure they are placed above the exact center of the dish. These pieces must be placed at the exact distance from the center of the dish. Look on the diagram for the right distance.

4. Bolt the dish to the pole. Tighten the bolts so that the dish doesn't wobble. Don't tighten them too much so that the dish can turn to track different birds.

5. Hook up motor, cables, and other parts. Be sure all connections are secure.

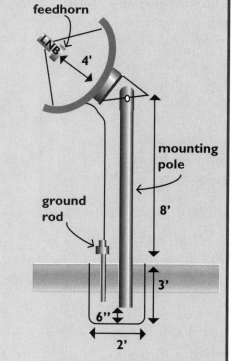

Choose the best answer to each question below. Base your answers on "Ground Mounting the Dish on a Pole" on page 72.

1. Where should the feedhorn and LNB be placed?

    a. 3–4 feet below the ground
    b. 8 feet above the ground
    c. 4 feet above the dish
    d. 2 feet below the ground

2. What should the installer do before the concrete gets hard?

    a. bolt the dish to the pole
    b. center the feedhorn above the dish
    c. center the LNB above the dish
    d. make sure that the pole is exactly plumb

3. Which component should be placed next to the mounting pole?

    a. motor
    b. ground rod
    c. cable
    d. feedhorn

4. How long is the mounting pole for this installation?

    a. 11 feet
    b. 8 feet
    c. 4 feet
    d. 3.5 feet

5. Why is it important for the pole to be secure and the bolts to be tight?

Check your answers on page 118.

Andrew began installing satellite dish systems in 1986. He had just finished high school.

Andrew liked science and math. He also loved watching television. He wanted to work in a job that was related to TV. He began working as a junior installer. He helped the senior installer, Mario, set up home satellite dish systems.

Back then, dishes were large and heavy. Mario needed Andrew's help to put together the petals of the dish. Together, they lifted the dish to the top of the mounting pole. Some dishes were more than 12 feet across.

After five years, Mario retired. Andrew became the senior installer. Andrew needed to train new junior installers.

Andrew showed the junior installers how to find the right spot to place the dish. He made sure they picked a spot that had a clear view to the south. He showed them how to mount the dish. Sometimes, it was set into concrete in the ground. Sometimes, it would sit on a flat concrete pad on top of the ground. Sometimes, the dish would be mounted on the customer's roof.

Andrew taught the new workers how to aim the dish. They learned how to figure out the right spot to direct the antenna. Their math and mechanical skills had to be exact to get the signal from 22,000 miles above the earth. If the dish's antenna was off by only 1 inch, the dish would not work right.

Finally, Andrew showed them how to test the system. He told them that pleasing the customer is important. He also made sure that the new installers were able to teach customers how to use their dish.

TALK ABOUT IT

1. Discuss how Andrew learned how to install dish systems.

2. Explain why it is important for dish installers to follow directions well.

Although satellite dishes may seem large and sturdy, they need good care. Dish installers not only set up new systems, but also maintain or care for older dishes.

Below are directions for caring for a dish system. Read the directions.

Caring for the Satellite Dish System

Taking good care of the dish and other installation pieces is very important. Harsh weather conditions can affect the system's parts. If parts are damaged or in bad working order, the system will not work properly.

Each dish should be inspected at least twice a year. Use the steps below when checking the system.

CHECKING SYSTEMS

Check the feedhorn and LNB. They must be directly above the center of the dish.

Check the mounting pole. It must be perfectly vertical.

Check the cables. Weather changes can damage the cables' casing. High temperatures can bake the casing. They may dry out and crack after a few years. Cold temperatures can also damage the casings. Replace the cables if they are damaged.

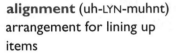

alignment (uh-LYN-muhnt) arrangement for lining up items

Check the bolts on the dish and the mount. Make sure they are tight. Loose bolts affect the dish's **alignment**. If the system has a movable dish, make sure the bolts are not too tight. The motor will burn out if they are too tight.

Keep the dish clean. Dirt, bird's nests, and other materials can interfere with reception. They change the surface of the dish. Snow and ice can also affect reception. It can weigh down the dish so that it is off track.

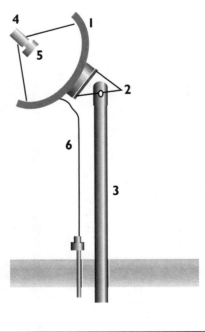

Key	
1. Dish	4. LNB
2. Mount	5. Feedhorn
3. Mounting Pole	6. Cable

Answer each question based on the directions on page 75.

1. What should the installer do if there is a cracked cable?

    a. tighten the cable
    b. clean the cable
    c. cut the cable
    d. replace the cable

2. Which two places on the diagram show where bolts should be checked?

    a. 1 and 2
    b. 3 and 4
    c. 5 and 6
    d. 4 and 5

3. What is at the top of the mounting pole in the diagram?

    a. the ground cable
    b. the feedhorn
    c. the mount
    d. the cables

4. Ice can change a system's reception. Why would it be important to keep the dish free of ice?

    a. Ice can weigh down the dish.
    b. Ice can loosen the bolts on the dish.
    c. Ice can freeze the satellite dish.
    d. Ice can shift the mounting pole.

5. According to the directions, when should a dish system be inspected?

    a. once a year
    b. at least twice a year
    c. in the fall
    d. in the spring

6. Which of the following is least likely to affect the system's reception?

    a. bird droppings
    b. snow
    c. heavy winds
    d. sunshine

7. What are two ways in which weather can affect how a system works?

Check your answers on page 118.

◆ LESSON WRAP-UP

In this lesson, you reviewed how to follow directions. Whenever you follow directions:

1. Read all of the directions and study all visual information, such as diagrams.
2. Make sure you know what the final result will be.
3. Think out each step in the directions.
4. Do the steps in order.
5. When you finish, check the final result.

Imagine that you are a satellite dish installer (SDI). You have almost finished installing a system for a customer. Answer the questions below. Give an example for each.

1. Why is it important for satellite dish installers to follow directions for installation?

2. Why is it important for satellite dish installers to follow directions for checking the systems?

Check your answers on page 118.

◆ UNIT THREE REVIEW

1. Imagine that you are responsible for a project. Something goes wrong. The project doesn't go as planned. You need to tell your supervisor what happened with the project and how you will correct it. Explain how you would use the skills in this unit to write a memo to her.

2. Imagine your friend is a computer service technician in another state. You ask him which type of computer you should buy. You also need to know where to buy it. He writes you a letter explaining everything. Which skills from this unit are you likely to use when reading his letter? Explain your answer.

3. Do you need to follow written directions at work? Do you also need to follow directions in diagrams? Write about a time at work when you had to use these skills.

Check your answers on page 118.

◆ Environmental ◆ Technology Occupations

In this unit, you will learn about environmental (ihn-vy-ruhn-MEHN-tl) technology workers. These workers are involved in cleaning our air, water, and soil. They also help businesses follow laws that are set up to keep our environment clean.

Workers in environmental jobs read business writing and technical material. This material includes reading on clean-up methods, safety notices, and training manuals.

This unit teaches the following reading skills:

◆ finding the main idea
◆ drawing conclusions
◆ classifying information

You will learn how workers in environmental technology use these reading skills in their work.

Working in Air Pollution Control

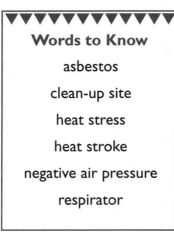

Words to Know

asbestos

clean-up site

heat stress

heat stroke

negative air pressure

respirator

Air pollution (puh-LOO-shuhn) means that the air we breathe is not as clean as it should be. Air pollution affects everyone. Air pollution control can protect the air.

Jobs in air pollution control require excellent reading skills. For example, air pollution control technicians must read manuals, instructions, memos, charts, graphs, and so on. When technicians are training, they often are told what information to look for when they read. When they are on the job, they need to know how to find important information for themselves.

The most basic reading skill on any job is **finding the main idea.** The main idea is the most important idea in what you are reading. Often, the main idea is stated at the beginning, middle, or end of a paragraph. The main idea also can be stated in a heading or a title of a reading. Sometimes, however, the main idea is not stated, and you have to figure it out from the details. Learning to look for main ideas when you read will help you become more successful on the job.

Job Focus

The job of an **air pollution control technician** is to measure and clean up air pollution. These technicians run tests in the field to find out if the air is polluted. When dangerous levels of pollution exist, technicians are sent out to remove the pollutants. They use special equipment to collect samples, record information, and conduct clean-up tasks. Some air pollution control technicians are specially trained in cleaning up one type of pollution, such as asbestos, lead, or various chemicals used in manufacturing.

Finding the Main Idea: How It Works

To **find the main idea** when you are reading, look for the most important point.

In longer readings, each paragraph usually has its own main idea. The main idea of each paragraph supports the main idea of the whole reading. Each paragraph also has supporting details that explain its main idea. Sometimes, a main idea is not stated. In these cases, use the details to figure out the main idea.

Read the paragraph below from a worker's instruction manual. Underline the main idea.

asbestos (as-BEHS-tuhs) fire-proofing material that contains fibers that are a health hazard

> *Instruction Manual* 3
>
> ## KEEP ASBESTOS OUT OF THE AIR
>
> One of the best ways to keep **asbestos** out of your lungs is to keep it out of the air. There are many ways to keep asbestos out of the air. These are given in the following basic rules, and you should memorize them.

Did you underline the words *keep asbestos out of the air?* If you did, you were correct.

Read the next paragraph that appears in the manual. Use the details to figure out its unstated main idea.

negative (NEHG-uh-tihv) **air pressure** (PREHSH-er) suction

> Follow these rules when you remove asbestos from a room or when you set up a job. Follow them when you clean up a work room. Remember these rules:
> 1. Keep the asbestos wet.
> 2. Contain the work area.
> 3. Filter the air.
> 4. Use **negative air pressure.**

What is the unstated main idea?

_____

_____

If you wrote something like *There are four basic rules for working with asbestos,* you are correct. This idea is the main idea. All the details are about this main idea.

clean-up site (KLEEN-up syt) place where pollution is being removed

When working on an asbestos **clean-up site,** workers are required to protect themselves from asbestos. An air pollution control technician has to have safe air to breathe. Below is a page from a training manual. It describes a protective device that is used in asbestos cleanup.

Read the page and think about its main ideas.

respirator (REHS-pih-ray-tawr) device worn over the nose and mouth to prevent breathing in harmful substances

---

16   Pollution Clean-up Manual

The Half-Mask, Air-Purifying Respirator

A half-mask, air-purifying **respirator** is the simplest respirator that can be used on an asbestos job. The bottom of the respirator's facepiece (the wide part) goes under your chin. The top of the facepiece (the narrow part) goes over your nose.

1.1   Purple Filters for Asbestos Clean-up

The half-mask is the least protective respirator allowed by law. It will not work unless the filter is made especially for asbestos. This filter is made up of two purple filters that catch the asbestos. They filter the air as you breathe.

1.2   The Protection Factor

Some respirators are better than others at keeping asbestos out. A half-mask, air-purifying respirator has a Protection Factor of 10. This means that for every 10 fibers in the air, one fiber leaks into the mask. For every 1,000 fibers in the air, 100 fibers leak in. Protection Factors range from 10 to 1,000.

1.3   Fit and Condition

The facepiece has to fit perfectly over your nose, cheeks, and chin. If it does not form an airtight seal, air and asbestos can leak in and around the edges of the mask. The air will not be filtered through the purple filters.

A respirator can't help you unless it's in perfect working order. Make sure all the parts are fastened and in good condition. Do this before you put it on. If you think there is anything wrong with your respirator, get it fixed.

LESSON 10 ◆ AIR POLLUTION CONTROL

Answer each question based on the training manual on page 82.

1. What is the main idea of the whole reading? Where does it appear?

2. What do the two supporting details in the first paragraph explain about the main idea?

3. What is the main idea of section 1.1?

4. What is the unstated main idea of section 1.2? (Remember, use details to figure it out.)

5. What is the unstated main idea of the second paragraph in section 1.3? (Remember, use details to figure it out.)

6. Explain how details helped you figure out the main idea of the whole reading.

7. Write a new title that states the main idea of the whole reading better.

Check your answers on page 119.

Danny is an air pollution control technician for the Department of Environmental Conservation (kahn-ser-VAY-shun). Danny's job is to test the air quality in public buildings. For the next few months, he is working on a big project at a public housing development. His work crew is testing to find out the level of lead in the air.

Danny wears a respirator and moves from floor to floor with an air pump. He pumps the air into an electronic device. It measures the air quality in various locations. Tests are done in the pipes, in the air shafts, and in the rooms. Danny follows a chart of each floor that tells him where to take air samples. On the chart, he writes down the date, time, specific location, and air-quality reading. He puts his initials on each line to show that he was the worker who took the reading.

Danny has been working in environmental technology for two years. He began as a technician's assistant (a clerk) in the county engineer's office. There, he assisted in laboratory work. He learned how to service the equipment and keep records on the computer. He helped prepare data used in the engineer's reports.

To get ahead in his career, Danny took some courses in a technical education program. He had to do a lot of technical reading and learn to use the equipment. He also learned how to prepare charts and graphs on the computer. On his first job, Danny read lab manuals and procedures manuals. He made a practice of reading the job reports. When the time came, he was able to take a civil service examination and qualify for his job as air pollution control technician.

TALK ABOUT IT

1. Why did Danny make a practice of reading the job reports in the laboratory?

2. Danny had to take an important test to qualify for his job. What are some ways you can prepare for a job examination?

Employees on technical jobs have to follow health and safety rules. Employers are required to provide equipment and training. They must also remind employees about how to keep safe and avoid problems on the job. Read the following memo about keeping safe in the heat. Then, answer the questions that follow.

heat stroke (heet strohk) severe illness caused when the body is overheated

heat stress (heet strehs) a milder form of heat stroke

DATE: Wednesday, July 8
  TO: Air Pollution Control Technicians
FROM: Health and Safety Office
  RE: Keeping Yourself Safe in the Heat

Hot weather can be a problem on the worksite during the summer months. You can protect yourself from health-related heat problems by following the guidelines below. Using caution can prevent lost work time due to **heat stroke** or **heat stress.**

1. Drink plenty of water. Your body loses lots of water when you sweat. It is best to drink some water every half hour. If not, take in 8 to 16 ounces of water at every break.

2. Eat foods that help your body hold water. Drink orange juice and eat bananas. Eat potato chips or another salty food once a day. But don't overdo salt intake. If you are on a low-salt diet, do not eat extra salt.

3. Take breaks. Your body will handle heat better if it can cool down sometimes. At least two breaks a day and a lunch break will help your body handle heat better.

4. Allow your body to get used to heat slowly. It takes about two weeks for your body to get used to working in the heat all day. New workers should work a half day in the heat for the first few days. If you are off the job for four days or more, allow time for your body to readjust to the heat when you come back.

5. Don't drink alcoholic beverages. Alcohol dries out your body. Even if you only have a small amount of alcohol at night, you are more likely to have problems with heat.

6. There are several warning signs of being overcome by heat to watch out for. They are:
   • feeling less alert
   • feeling less coordinated
   • getting a headache
   • feeling sick to your stomach.

Remember, your good health is our concern!

FINDING THE MAIN IDEA

CHECK YOUR UNDERSTANDING

Answer each question based on the memo on page 85. Make an X on the line or lines next to the correct answers to each question.

1. What is the main idea of the whole memo?

___ a. Air pollution causes heat stroke.

___ b. You can take steps to prevent problems from heat.

___ c. Salt is not good if you are on a low-salt diet.

___ d. The company is responsible for workers' health.

2. What is the main idea of guideline 2?

___ a. Drink some orange juice and eat bananas.

___ b. Eat potato chips or another salty food once a day.

___ c. Eat foods that help your body hold water.

___ d. The body may need extra salt.

3. Which details support the main idea of guideline 4?

___ a. It takes the body two weeks to get used to working all day in the heat.

___ b. Allow your body to get used to heat slowly.

___ c. New workers should work half a day at first.

___ d. If you're off the job for four days or more, allow time for your body to readjust to the heat.

4. What is the main idea of guideline 5?

___ a. Alcohol makes heat harder on the body.

___ b. Don't drink alcoholic beverages.

___ c. Drink only beer but nothing stronger.

___ d. The company doesn't allow drinking.

5. Your boss has asked you to write a summary of the memo to post at the worksite. Begin by stating the main idea of the whole memo. Then, write the main idea of each paragraph. Write your summary below.

Check your answers on page 119.

◆ LESSON WRAP-UP

In this lesson, you learned how to find the main idea when you read. The main idea may be stated in a title, a heading, or in the first paragraph of a reading. The main idea may also be unstated. Use supporting details in each paragraph to figure out an unstated main idea.

When you read, follow these guidelines:

1. Look for the main idea of the passage in the first sentence of the first paragraph or in one or more other sentences.
2. For longer pieces, read each paragraph, and look for the main idea in each one.
3. If the main idea of the entire passage is not stated, figure it out from the details. Remember, the main idea of a paragraph can be a detail that supports the main idea of the entire passage.

Think of the types of reading materials you use on your job or the type of materials you read most often at home. Then, complete the sentences below.

1. Three places that I look for the main ideas when I read are:

   a.

   b.

   c.

2. The skills I learned in this lesson that will help me improve my ability to find the main idea when I read are:

Check you answers on page 119.

Working in Waste Management

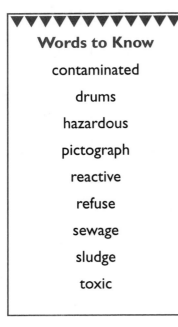

Words to Know

contaminated

drums

hazardous

pictograph

reactive

refuse

sewage

sludge

toxic

Workers in waste management sometimes handle harmful materials. Often, training for a job in waste management includes reading educational materials on safety rules. On the job, workers also read signs, labels, and instructions about how to complete a job safely.

Waste management workers should be able to make good decisions or judgments based on the information they read. This is called **drawing conclusions.**

To draw conclusions, you look at all the given details or facts. Then, you think about what you already know, and make a decision or judgment. Practicing this skill will help prepare you for a job in waste management technology.

Sometimes, you will have plenty of time to draw your conclusions. Other times, you may face a situation that you did not expect. Being able to draw conclusions quickly and correctly from your reading will help you become a successful worker.

Job Focus

Waste management technicians work with material that has been thrown away, or *discarded.* Some workers sort, label, and group material at a transfer station. They may wear special clothing, including masks. Others go from one place to another to pick up material for disposal. The job may require special equipment and the handling of metal containers.

Jobs in waste management are increasing as more materials are being *source separated.* This means that waste materials are separated into containers for medical, recyclable, and harmful waste. Then, workers discard them, following legal standards set up for each type of waste.

Drawing Conclusions: How It Works

When you **draw conclusions** from what you read, you think about what the reading says. You connect new information to information that you already know. Then, you are able to make a decision or judgment.

Once you have drawn a conclusion, you can test it by asking yourself questions. Some questions you can ask are:

- Is my conclusion logical—does it make sense?
- Is it based on the information given to me?
- Have I carefully read and thought about all the information?
- Have I added information of my own that changes the facts?

Drawing conclusions is very important when you read information about how to do your job. When you are given something to read, take the time to read all of it. You may need to read it more than once. Paying careful attention to details is important. This may help you draw correct conclusions based on the facts.

Below is part of an information sheet about waste. It has definitions about different kinds of waste. Read the definitions.

sewage (SOO-ihj) waste from sinks, bathtubs, and toilets; this waste is carried by water through pipes from buildings

hazardous (HAZ-uhr-duhs) harmful or dangerous

1. Solid waste includes household trash, **sewage,** and industrial materials.
2. The Environmental Protection Agency (EPA) defines **hazardous** waste as a type of solid waste.
3. Hazardous waste is solid waste material that is a threat to human health or the environment.

Based on the definitions above, which of the following conclusions can you draw?

Conclusion 1: Some waste material is hazardous; some is not.
Conclusion 2: All solid waste material is hazardous.

If you chose the first conclusion, *Some waste material is hazardous; some is not,* you are correct. Definition 2 tells you that hazardous wastes are one type of solid waste. Definition 3 tells you that hazardous waste is a threat to health. Therefore, the conclusion that some solid waste is not hazardous is correct.

As you just learned, some waste material is hazardous and some is not. Waste management technicians will need to know the difference. Read the following page from a training manual. Then, answer the questions that follow.

Solid and Hazardous Waste

Solid waste refers to any garbage, trash, or **refuse**. It also includes **sludge** from waste treatment plants, and air pollution control facilities. Solid waste is not always found in solid form. It may be a liquid, semisolid, or gas. These types of solid waste usually come from industrial, commercial, mining, and agricultural operations.

Hazardous wastes are a type of solid waste. They can be solids, liquids, or gases. They may appear in a variety of places and in many forms: in **drums**; in pits or ponds; in sludge; as part of **contaminated** soil; in bottles or other breakable containers; in aboveground or underground storage tanks; and as part of some building materials.

Lists of hazardous materials are available from the EPA (Environmental Protection Agency). Make sure that the list at your worksite is up to date, because the EPA often makes changes to it.

refuse (REH-fyooz) trash or garbage; material that is useless

sludge (sluj) solid matter produced by water and sewage treatment

drums (drumz) metal containers or barrels used for storing

contaminated (kuhn-TAM-ih-nayt-ehd) unfit for use; impure

CHECK YOUR UNDERSTANDING

Write **T** in the blank if a conclusion is true or **F** if it is false. Some conclusions are based on information that is not stated in the reading. Write **N** next to these statements.

___ **1.** Sludge may or may not be a hazardous waste.

___ **2.** Waste from mining operations is always hazardous.

___ **3.** Cyanide is a hazardous material.

___ **4.** Hazardous waste in liquid form may be found in a container.

___ **5.** Hazardous waste is never found in water.

___ **6.** Contaminated soil can be hazardous.

___ **7.** Household garbage is not usually hazardous.

___ **8.** Most hazardous waste comes from industry.

___ **9.** The EPA never updates its hazardous waste list.

Check your answers on page 119.

Joyce works at the South Township Waste Management Facility. She is a waste management technician at this waste station. She makes sure that waste material is properly stored for disposal. The material is separated into large dumpsters. The station handles household trash, garbage, and recyclable material. Joyce uses special equipment to transfer this material to smaller containers of different colors. She also handles hazardous waste material. This material comes to the station already packed in red containers. Joyce prints out labels on the computer and places them on the containers. She counts the containers and records how many containers of each kind are being sent out.

Joyce says, "I'm not afraid to handle hazardous material. It is carefully packed in containers and labeled. In the training program, I learned about all types of materials that come to the station. I read instruction manuals on how to do things right to avoid mistakes."

Joyce works as part of a team. She wears protective clothing, including a plastic coat, cap, and gloves. She also places a small mask over her mouth so that she doesn't breathe in contaminated materials. "On the job, we have sheets of rules to follow. Everything is labeled and written down. The supervisor checks what we do. It's like working in any job where something could go wrong. Everyone has to stay alert, be careful, and follow instructions."

TALK ABOUT IT

1. Many jobs involve some risk. Name some jobs in which workers have to avoid harm to themselves or to others.

2. Would you want to have a job that could cause harm if you made a mistake? Discuss why or why not.

Waste is treated, stored, or disposed of at a waste disposal site. At a waste disposal site, workers do not do much reading of instructions and manuals. They are more likely to read notices and signs posted there. They may also read labels on containers. Read the following notice that was given to a work crew before going out on a job. It explains the labeling on waste containers.

N O T I C E

TO: All Handlers
RE: Job #14072 Department of Fire Protection
           Job #14073 Department of Transportation
DATE: Monday, March 1

Instructions:
Drums from these jobs will be taken to storage. Drums will be loaded onto hand trucks. Each kind of waste will be moved to its own area. Make sure crews enter information correctly on the field computer.

Labels:
Most labels show the name of the material. Some have a written description. If the material is hazardous, a warning will be written on the label. Department of Fire Protection uses a diamond-shaped **pictograph** to show contents. Look at the picture at right. Notice that four sections of the diamond use patterns to show contents. Here is what each pattern means:

   1) A dotted upper-left section means health hazard.
   2) A striped lower-left section means fire hazard.
   3) A black upper-right section means **toxic** material.
   4) A checkered lower-right section means any other type of material, such as chemicals or **reactive** materials.
   5) White sections mean non-hazardous material.

Caution: SOME CONTAINERS MAY NOT CONTAIN THE SUBSTANCE ON THE LABEL. Some drums have been reused. All handlers must use good judgment when checking to make sure labels are correct. If necessary, open the drum. Make a visual or chemical test of drums to make sure waste is what the label says. Make sure that you use the standards chart available in the office.

---

pictograph (PIHK-toh-graf) drawing that has a standard meaning, such as a circle with a line across it that means *no* or *don't*

toxic (TAHKS-sihk) poisonous

reactive (ree-AK-tihv) able to explode

For each item, circle the best conclusion that can be drawn. Base your answers on the information in the notice on page 92.

1. What type of information is probably on the standards chart referred to in the last paragraph? (Circle *two*.)

    a. cost of handling various types of waste
    b. taste, appearance, and smell of hazardous materials
    c. color, chemical makeup, and smell of waste materials
    d. simple tests that can check what the material is

2. Why would the Department of Fire Protection use a pictograph? (Circle *two*.)

    a. Someone in the department likes to draw.
    b. A picture attracts attention more than words do.
    c. A lot of information can be stated in a small space.
    d. none of the above

3. Information about the drums must be put into the computer to

    a. keep track of each job.
    b. record in which area of the facility each drum is stored.
    c. monitor any hazardous substances.
    d. all of the above

4. You can conclude from this memo that

    a. the company wants workers to follow standard procedures.
    b. managers do not trust handlers' judgment.
    c. this facility can handle only one job at a time.
    d. material is stored in various size drums.

5. How do workers keep track of containers? (Circle *two*.)

    a. by marking them with warning labels
    b. by storing them in the correct area
    c. by entering them on the field computer
    d. by checking the standards chart

6. This Department of Fire Protection label means the container holds

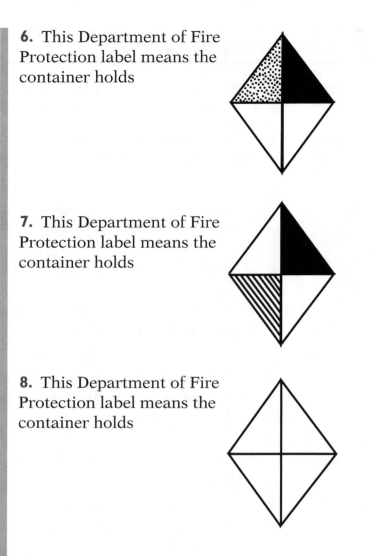

7. This Department of Fire Protection label means the container holds

8. This Department of Fire Protection label means the container holds

Check your answers on page 120.

◆ LESSON WRAP-UP

In this lesson, you learned how to draw conclusions. To draw correct conclusions, you need to follow a certain thinking process. This process is made up of the following steps:

1. Read the information.
2. Look for main ideas.
3. Pay attention to details.
4. Connect the information to what you already know.
5. Make judgments or decisions based on the material.
6. Test your conclusions by asking yourself questions.

Some questions to ask yourself are:

- Is my conclusion logical—does it make sense?
- Is it based on the information given to me?
- Have I carefully read and thought about all the information?
- Have I added information of my own that changes the facts?

You also learned that you may need to draw conclusions quickly at times. Each time you read, practice the thought process you have learned here. Practice will help you become skilled at drawing correct conclusions.

Read each paragraph. Then, answer the questions that follow.

1. You keep your protective work clothing in your locker at work. A friend calls you over the weekend and says his company needs an extra worker for a waste disposal job on Sunday. You don't have a key to the locker room. Your friend says, "You may only need to wear gloves. I don't think we'll be handling any contaminated material."

  a. What conclusion can you draw from this situation?

  b. List the facts and information you used to come to this conclusion.

2. It is not always possible to draw the right conclusion from information given. When you make a mistake, you may wonder, "Is it my fault? Is it someone else's fault?" List some reasons that you may draw incorrect conclusions when you read. Then, list ways you can try to avoid this problem.

Check your answers on page 120.

Working in Recycling Jobs

Words to Know

▼▼▼▼▼▼▼▼▼▼▼

degrade

natural forests

non-renewable resource

renewable resource

tree farms

Recycling is the practice of using waste material again, in another form. Things people have thrown away—paper, plastic, tin, glass, and food products—can be reused. They can be made into new and different products. Recycling saves energy and resources. Technology helps people and companies recycle goods.

Most companies that produce products have recycling programs. They use recycled materials to make new products. They also put their products in packaging that can be recycled.

Consumers—people who buy products—also play a major role in recycling. They can choose to buy products that come in packaging that can be recycled. They can also recycle their used goods.

When recycling workers read on the job, they sometimes **classify information**. To *classify* means "to organize facts and details into like or related groups." Grouping items that are alike in some way helps you understand information.

Job Focus

More than 6,000 U.S. communities have recycling programs. There are many types of jobs for **recycling specialists.** Collection centers and processing plants hire workers to handle material as it comes into the centers. These workers sort the material, put it in containers, and label it. Businesses hire employees to market and sell products made from recycled materials. Manufacturing companies hire technicians to take materials through recycling processes. People with good technology skills can use their skills on recycling jobs.

Classifying Information: How It Works

When you **classify information,** you separate it into groups. You form the groups by putting similar items together. Suppose you work as a paper recycling technician. You examine used paper products that come into the recycling plant. You look for items in the paper that need to be removed. Your boss has asked you to make a list of the unwanted objects found in used paper. The plant is thinking about buying equipment to screen out unwanted objects.

Over a week, you take notes and list these items:

metal paper clips	vinyl binders
plastic paper clips	rope
metal clip binders	staples
string/twine	cloth binders
rubber bands	pens/pencils

Before giving the list to your boss, you decide to classify the objects. You group them based on the material they are made of. Then you make the following report.

Here is a classification of unwanted objects in paper for recycling:

Metal: paper clips, staples, binders
Plastic: paper clips, pens, binders
Wood: pencils
Rubber: rubber bands
Cloth: binders, string, twine, rope

With this report, your boss can see the kinds of materials that the special equipment will have to handle.

Can you think of another way you could have classified the objects in this list? How about using two groups—hard items and soft items? Try classifying the items according to these two groups.

Hard: _____

Soft: _____

Did your groups look like this? Hard: *metal paper clips, plastic paper clips, metal clip binders, vinyl binders, staples, cloth binders, pens, pencils.* Soft: *string, twine, rubber bands, rope.*

As you can see, information can be classified in different ways. How you group depends on how you want to use the information.

Different kinds of paper are recycled to make different kinds of products. Some examples of products made from recycled paper are greeting cards, stationery, paper towels, notebook binders, and wrapping paper.

A paper company wants to recycle its waste paper. Jonathan Davis is a technician at the company. His boss wrote him the memo shown below. It lists all the kinds of waste paper that the business throws away. He asked Jonathan to review the list and suggest ways that the paper can be recycled. Read the memo.

DATE: January 5
TO: Jonathan Davis
FROM: Edgar Barton
SUBJECT: Paper Recycling

Here is a list of paper products that are discarded in a typical week. We want to recycle this paper.

Please review the list below. Let me know the items that we (1) can handle here, (2) will need to send to the North Shore plant, and (3) cannot recycle.

plain white paper	paper with labels
plain colored paper	newspapers
colored folders	magazines
envelopes with labels	glossy brochures
phone books—	envelopes
White Pages	without labels
phone books—	sticky notes
Yellow Pages	fax machine
junk mail letters	paper

Thanks for your help!

CHECK YOUR UNDERSTANDING

Answer each question based on the memo on page 98.

1. You know that your plant recycles only white paper. The North Shore plant recycles white and colored paper. Neither plant can use items that contain any type of glue or sticky substance. Classify the items in the list for your boss.

Our Plant

1.   2.

3.

North Shore Plant

4.   5.

6.   7.

8.   9.

10.

Can't Use

11.   12.

13.   14.

2. Before the paper is sent out for recycling, you need to weigh it on weighing scales. Classify the items according to size for weighing.

Small

1.   2.

3.

Medium

4.   5.

6.   7.

8.   9.

10.

Large

11.   12.

13.   14.

Check your answers on page 120.

Cora works in the paper-processing department of the Johns Brothers Paper Mills. Each week, several tons of used paper products come to the mill for recycling. Technicians put the paper into a machine called a *hydropulper.* This machine works like a high-powered blender. It crushes and grinds the paper and mixes in water. The mixture turns into a thick mass, with a texture like oatmeal. A chemical process removes the ink. To remove items such as paper clips and staples, machines pour the paper mixture through screens. Then, the paper is dried and rolled out into large sheets.

To learn to use the machines, Cora had to read operations manuals. She had to learn how to work the equipment that loads paper into the machine at the beginning of the process. She helps the engineer in the chemical-processing stage. And she has to reset the equipment that dries and rolls the paper to make various kinds of paper. The recycled paper is used to make many different kinds of paper products.

Cora enjoys her job. She knows that recycling reduces garbage and pollution. She talks to her friends about being smart consumers. For example, she tells them to buy only products that are put in packaging that can be recycled. Cora knows that each person can make a difference. Just the simple act of throwing a soda can into a recycling bin helps make our environment cleaner and more attractive.

TALK ABOUT IT

1. Name at least two more recycling tips that Cora might give her friends.

2. What do you do to keep our environment clean? Make a chart with two columns labeled "Things I Do Now" and "Things I Can Do in the Future." Use a separate sheet of paper. Show your chart to a classmate, and exchange ideas.

Your company is trying to encourage stores to use paper bags instead of plastic bags. Some people feel that paper is better for the environment. Others feel strongly that using plastic is better. Your company is sending a letter to area stores about this issue. Read the letter below.

Monday, September 29

Dear Environmentally Concerned Merchant:

As you know, people disagree about whether using paper bags or plastic bags helps the environment. As a merchant, you have to think about costs. However, I am writing to ask that you also think about how your decision affects the environment. Below you will find good reasons for using both paper and plastic. This information will help you make an informed decision.

Paper is a **renewable resource**. Paper manufacturing does not kill **natural forests** anymore. Much of the paper we use today is made from trees planted and harvested on **tree farms.** Paper grocery bags contain about 20 percent to 35 percent recycled fibers. And there is a large industry in place for recycling bags. According to a survey, 21 percent of communities accept brown paper bags for recycling. Only about 4 percent accept plastic bags.

The argument for plastic bags is that they are reusable. People use them for lining wastebaskets. They can also be used to carry anything and everything. It's true that plastic lasts longer than paper, but paper bags do not **degrade** any faster than plastic. It's also true that plastic weighs less and is less bulky to store than paper. The downside is that plastic production uses a **non-renewable resource**—oil. Also, plastic bags that get into streams, rivers, and oceans are hazardous to fish and other living things.

We advise you to make a decision based on the recycling facilities you have in your area. And that's where we come in! We recycle paper. We're here to serve you and your customers. Call us at 555-2795 to discuss our recycling program.

Sincerely yours,

Joel Coswell, Manager

---

renewable resource (rih-NOO-uh-bl REE-sawrs) resource, such as water, that is never really used up

natural forests (NACH-er-uhl FAWR-ihsts) forests that renew themselves naturally

tree farms businesses that grow trees for the purpose of making paper

degrade (dih-GRAYD) to reduce back to its elements

non-renewable resource (nahn-rih-NOO-uh-bl REE-sawrs) resource that cannot be replaced once it is used up

Classify the information in the letter on page 101. Use the categories below.

Points for Using Paper

Points Against Using Paper

Points for Using Plastic

Points Against Using Plastic

Check your answers on page 121.

◆ LESSON WRAP-UP

In this lesson, you learned how to organize information by classifying it. This means putting the information into groups. The groups should make sense, based on how you need to use the information.

When you classify information, you can understand it more easily. You can find details quickly and understand where they fit. Classifying information helps you when you need to make decisions. Sometimes, when you divide information into categories, you may want to make a decision about each group. Classifying also helps you compare categories, such as when you look at the points for and points against something. Use the information that follows to practice categorizing and classifying.

The three R's of recycling are **R**esist, **R**educe waste, and **R**euse. Classify each action listed below into one of these three categories. Place the letter of the item under the **R** that fits. The first one has been done for you.

a. Buy large quantities (laundry soap, for instance).
b. Take a coffee mug to work instead of using paper cups.
c. Don't buy things that you don't need.
d. Don't buy single-serving packages of food.
e. Save glass jars and plastic containers for leftovers.
f. Don't buy things packaged in material that can't be recycled.
g. Try not to buy take-out foods in plastic containers.
h. Use plastic bags many times.

Resist

1.

2.

3.

Reduce

1. a

2.

Reuse

1.

2.

3.

Check your answers on page 121.

◆ UNIT FOUR REVIEW

1. Give the steps for finding the main idea in a long passage.

2. Explain the steps that you should use when you draw conclusions.

3. How does classifying information help you use it?

4. Describe how the three reading skills you practiced in this unit will help you be a better reader on the job.

Check your answers on page 121.

RESPELLING GUIDE

Use the following guide to help you pronounce long and hard words.

Sound	Respelling	Example of Respelling
a as in hat	a	hat
a as in day, date, paid	ay	day, dayt, payd
vowels as in far, on, bother, hot	ah	fahr, ahn, BAH-thuhr, haht
vowels as in dare, air, heir	ai	dair, air, air
vowels as in saw, call, pore, door	aw	saw, kawl, pawr, dawr
e as in pet, debt	eh	peht, deht
e as in seat, beef, chief, **y** as in beauty	ee	seet, beef, cheef BYOO-tee
vowels as in learn, urn, fur, sir	er	lern, ern, fer, ser
i as in sit, bitter, **ee** as in been	ih	siht, BIHT-uhr, bihn
i as in mile, **ei** as in height	eye	meyel, heyet
o as in go	oh	goh
vowels as in boil, toy	oi	boil, toi
vowels as in how, out, bough	ow	how, owt, bow
vowels as in up, come	u	up, kum
vowels as in use, use, bureau, few	yoo	yooz, yoose, BYOO-roh, fyoo
vowels as in look, put, foot	oo	look, poot, foot
vowels as in bitter, action	uh	BIHT-uhr, AK-shuhn

Consonants are respelled as they sound. Here are a few examples.

c as in cat	k	kat
c as in dance	s	dans
ch as in Christmas	k	KRIHS-muhs
g as in gem	j	jehm
s as in laser	z	LAY-zuhr
ph as in phone	f	fohn

R E S O U R C E S

The following organizations and publications may provide more information about the jobs covered in this book.

United States Government
U.S. Department of Labor, Employment and Training Administration

Adult Training Programs include the following:
Job Training Partnership Act (JTPA)
This program provides job training for disadvantaged adults who face significant employment barriers. For more information, write:

> Office of Employment and Training
> Programs, Room N4469
> U.S. Department of Labor
> 200 Constitution Ave, N.W.
> Washington, DC 20210

ON THE INTERNET: http://www.doleta.gov/programs/programs.htm

Apprenticeship Training
The Bureau of Apprenticeship and Training registers apprenticeship programs in 23 states. It also assists State Apprenticeship Councils in 27 states, the District of Columbia, Puerto Rico, and the U.S. Virgin Islands. For further information, write or call:

> Bureau of Apprenticeship and Training
> U.S. Department of Labor
> 200 Constitution Ave, N.W.
> Washington, DC 20210
> (202) 219-5921

ON THE INTERNET: http://www.doleta.gov/programs/programs.htm

The Bureau of Labor Statistics has descriptions of working conditions for a wide variety of specific occupational areas.

For more information on the Bureau's publications, write to:

> Bureau of Labor Statistics
> Division of Information Services
> 2 Massachusetts Avenue, N.E.
> Room 2860
> Washington, DC 20212

Information specialists provide a variety of services by telephone: (202)606-5886
To send a question by fax, please call (202)606-7890
ON THE INTERNET: http://stats.bls.gov

For general career information and a directory of accredited private career and technical schools offering programs in the job areas covered by this book, write to:

> Accrediting Commission of
> Career Schools
> 2101 Wilson Blvd.
> Suite 302
> Arlington, VA 22201

Unit 1 LASER TECHNOLOGY OCCUPATIONS

> Laser Institute of America
> 12424 Research Parkway
> Suite 125
> Orlando, FL 32826

The Laser Institute of America provides laser safety information and training. Call for more information:
1-800-34-LASER(345-2737).
PHONE: (407) 380-1553
ON THE INTERNET:
http://www.creol.ucf.edu/~lia/

> Institute for Interconnecting
> and Packaging Electronic Circuits

2215 Sanders Road
Northbrook, IL 60062-6135
PHONE: (847) 509-9700
FAX: (847) 509-9798
ON THE INTERNET:
http://www.ipc.org/index.html
The Institute for Interconnecting and
Packaging Electronic Circuits provides
video training for printed circuit
fabrication and electronic assembly.

Periodicals

LIA Today
The official newsletter of Laser
Institute of America

Laser Focus World
PennWell Publishing Co., Tulsa, OK
ON THE INTERNET:
http://www.lfw.com/www/home.htm

Unit 2 JOBS IN COMPUTERS AND ELECTRONICS

The International Society of Certified
Electronics Technicians (ISCET)
2708 West Berry St.
Fort Worth, TX 76109
PHONE (817) 921-9101
FAX (817) 921-3741
ON THE INTERNET:
http://www.iscet.org/iscet.htm
Electronic Industries Foundation
919 18th Street NW, Suite 900
Washington, DC 20006
PHONE: (703) 907-7500
FAX: (202) 955-5837
EMAIL: eifcorp@aol.com

Desktop Publishers Journal
(DTP Journal)
462 Boston St.
Topsfield, MA 01983-1232
PHONE: (508) 887-7900

FAX: (508) 887-6117
ON THE INTERNET:
http://www.dtpjournal.com/

UNIT 3 COMMUNICATIONS TECHNOLOGY OCCUPATIONS

Electronics Technicians Association,
International, Inc.
602 N Jackson
Greencastle, IN 46135
PHONE: (317) 653-4301
ON THE INTERNET:
http://www2.fwi.com/~n9pdt/eta.html

International Mobile
Telecommunications Association
(IMTA)
1150 18th Street, NW, Suite 250
Washington, DC 20036
TELEPHONE: (202) 331-7773
FAX: (202) 331-9062
ON THE INTERNET:
http://villagenet.com/~imta/

Schools for technicians
Certified Satellite Installer C.S.I. write to:
Satellite Dealers Association, Inc
602 N Jackson
Greencastle, IN 46135
PHONE: (317) 653-4301
FAX (317) 653-8262

Photo Marketing magazine
Photo Marketing Association
International
3000 Picture Place
Jackson, MI 49201
PHONE: (517) 788-8100
FAX: (517) 788-8371
ON THE INTERNET:
http://www.pmai.org/index.html

Photo Imaging Education Association
Association of Professional Color

Imagers (APCI)
sections of Photo Marketing Association
International

Unit 4 ENVIRONMENTAL TECHNOLOGY OCCUPATIONS

American Waterworks Association
6666 West Quincy Ave.
Denver, CO 80235

Periodicals
American Waste Digest
226 King Street
Pottstown, PA 19464
Toll Free: 1-800-442-4215
Phone: (610) 326-9480
Fax: (610) 326-9752
Email: awd@becnet.com
On The Internet:
http://www.americanwastedigest.com/

BioCycle
Journal of Composting & Recycling
419 State Ave.,
Emmaus, PA 18049
Phone: (610) 967-4135
Fax: (610) 967-1345

Recycler's World
A world wide site for information related
to recyclable items or materials.
On The Internet:
http://www.recycle.net/recycle/index.html

Solid Waste Association
  of North America
P.O.Box 7219
Silver Spring MD
USA 20910-7219
Phone: (301) 585-2898
Fax: (301) 589-7068

National Solid Wastes
  Management Association
4301 Connecticut Ave. N.W., # 300
Washington, DC 20008
Phone: (202) 244-4700
Fax: (202) 966-4841

National Recycling Coalition (USA)
1727 King Street, Suite 105
Alexandria, VA 22314
Phone: (703) 683-9025
Fax: (703) 683-9026

G L O S S A R Y

adapter device for connecting parts that otherwise would not fit

alignment arrangement for lining up items correctly

asbestos fire-proofing material that contains fibers that are a health hazard

assemblers workers who put together equipment

bills of materials (BOMs) lists of all the parts needed for a specific job

bonus extra money added to a regular salary or check

bunny suits special suits that cover and protect the body

capacitor an electrical part that stores a charge for a short time

cellular phones telephones that operate on radio waves

clean-up site place where pollution is being removed

components parts of a system

conector a part that joins other parts

contaminated unfit for use; impure

contamination pollution or dirt

continuing education classes for adults to help them on the job or in their personal lives

custom prints photographs developed in a special way for a customer

degrade to reduce back to its elements

developer chemicals used to make the image appear in film

drums metal containers or barrels used for storing

enlarger machine that projects a light through a lens to resize an image

ergonomics field of safe and healthy work practices

equator imaginary circle that goes around the exact middle of the earth

exposure time number of seconds needed for light to pass through a negative and onto special paper

feedhorn part of a satellite dish system that collects signals that bounce off the dish

fixer chemical used to clear off unused chemicals when developing film

flexible schedules work hours for each employee that can vary from set work times

focusing adjusting the lens of a camera to make the subject clearer

font style for printing letters, numbers, and symbols

gain-sharing sharing of company profits with employees

graphics pictures or other visuals

ground rod part of a satellite dish system that protects it from lightning strikes and electrical problems

hazardous harmful or dangerous

heat stress mild form of heat stroke

heat stroke severe illness caused when the body is overheated

human resources department in a company responsible for employee hiring and benefits

inspection checking of parts to see if

they are good parts or scraps

keyboard part on the computer with keys for typing

laser machine that produces a steady, powerful beam of light

LNB part of a satellite dish system that helps process the signals

logic chip an electrical part that uses math to find a result

logo symbol that a company or group uses to identify itself

microphone a device that changes sound into electricity

models different styles or types of a product

monitor screen on the computer where you see your work

natural forests forests that renew themselves naturally

negative air pressure suction

negatives developed film

non-renewable resource natural resource that cannot be replaced once it is used up

obstructions objects that block other objects

ordinances local government laws or rules

packing list record of items in a shipment

particles very tiny pieces

pictograph drawing that has a standard meaning, such as a circle with a line across it that means *no* or *don't*

plumb place straight up and down and side to side

policies general practices, guidelines or standards

portable able to be carried easily

precision lenses high quality, carefully shaped pieces of curved glass

procedure course of action

product ID number tracking number

productivity keeping costs and waste low, but output high

purchase order list of items ordered by a customer

quality control method used by companies to produce better products

reactive able to explode

readability ease or difficulty of reading material

refuse trash or garbage; material that is useless

renewable resource resource, such as water, that never really is used up

resistor an electrical part that opposes the passing of an electric current

respirator device worn over the nose and mouth to prevent breathing in harmful substances

scalpels small knives with sharp blades used in surgery

scanners machines that use beams of light to read type and images by comparing the dark and light spaces on the item

scrap damaged parts or discarded products

sensors parts of equipment that pick up, or "sense," messages or information

sewage wast from sinks, bathtubs, and

toilets; this waste is carried by water through pipes from buildings

shutter part of the camera that opens and shuts to let light in or keep it out

sludge solid matter produced by water and sewage treatment processes

stop bath chemicals used to stop the development of the image

supervisor person who manages other employees or workers

talk time amount of time that you can talk on a mobile phone without having to recharge the battery

termination ending, as in separation from a job

toxic poisonous

track follow each step in a process

tree farms businesses that grow trees for the purpose of making paper

tripod three-legged support used to hold a camera

violation breaking of a rule

visor movable brim attached to a hood or a hat

watts units of electrical power

INDEX

A N S W E R   K E Y

UNIT 1: LASER TECHNOLOGY OCCUPATIONS

Lesson 1: Working as an Assembler
CHECK YOUR UNDERSTANDING
page 5
1. c
2. b
3. b
4. a
5. d

CHECK YOUR UNDERSTANDING
page 8
1. b
2. d
3. a
4. c
5. Your answer may be similar to this:
LaserVu builds, sells, and services many kinds of industrial lasers.

LESSON WRAP-UP
page 9
Your answers may be similar to these:
1. I read an article in a magazine.
2. The topic of the piece is United States national parks.
3. The main idea of the piece is that U.S. national parks are beautiful and exciting places to visit.
4. Understanding the main idea of the piece helped me to see how valuable these parks are. They hold some of our nation's most beautiful treasures. I never thought much about nature before reading this article. Now, I think I would like to go see some of the parks.

Lesson 2: Working as a Laser Technician
CHECK YOUR UNDERSTANDING
page 13
1. b, c
2. d, g

3. a, e, f
4. a. The company hands out its drug-free workplace policy statement every year.
   b. The company posts the statement on three bulletin boards.
   c. The company includes a copy of the policy in the employee manual.
5. The employee can be fired, or terminated
6. Your answer may be similar to this:
Yes. I think it's important that employees don't use drugs. They must be completely alert when using the equipment. If they aren't alert, they may hurt themselves or someone else.

CHECK YOUR UNDERSTANDING
page 16
1. dressing area
2. bunny suit, hood, and slippers
3. head, eyes, body, feet (Any three are correct.)
4. dirt, dust particles, and skin particles
5. visor or safety goggles
6. F
7. F
8. T
9. F
10. Your answer may be similar to this:
It should cover your entire body. It should fit comfortably. You should be able to move easily in it.
11. Your answer may be similar to this:
It sounds like you are working alone. It seems you must take your time and be very careful. You would probably hear the hum of the machines controlling the air and temperature.

LESSON WRAP-UP
page 17
Your answer may be similar to this:
Dear Sarah,
   I am sorry about the accident that

happened last week.

I was backing my car out of the driveway next to our apartment building. I didn't see the grocery bags on the driveway where your nephew had left them. I ran over them before I realized what happened.

I got out of the car and apologized to your nephew. I didn't wait to talk to you because I was in a hurry to go on vacation. But I haven't been able to relax because of what happened before I left, so I wrote you this note.

In the future, I'll look in the driveway before I back up. And, as I told your nephew, I gladly pay you back for the groceries. Sorry.

Sincerely,
Sam Chancellor

Lesson 3: Working as a Laser Operator

CHECK YOUR UNDERSTANDING

page 21

1. b
2. e
3. b
4.

Laser #	Good Parts to Scrap		Good Parts/Scrap
1	243 to 13	or	243/13
3	503 to 52	or	503/52
5	102 to 4	or	102/4
7	430 to 26	or	430/26
9	489 to 17	or	489/17

Your answer may be similar to this: Rita controlled the quality of her batch better. Even though she had more pieces to process, she damaged fewer pieces than Tyrone did. When you compare the number of scrap pieces to the total batch, Rita's percentage of scrap is lower.

CHECK YOUR UNDERSTANDING

page 24

1. b
2. c
3. d
4. c
5. Yes, they did earn it. The ceramics

department outperformed the company in February, April, May, and June.

LESSON WRAP-UP

page 25

A possible answer might look like this:

Week	Grocery Bill
1	$125.00
2	$75.00
3	$100.00
4	$95.00

UNIT ONE REVIEW

page 26

Your answers may be similar to these.

1. My co-worker was in a big hurry. I needed to tell her the main point of the memo quickly. I think I would use the skill of finding the main idea. I would try to remember the most important point. It would help me give her a broad understanding of the whole memo.

2. When my co-worker had more time, I could go into more detail. We could talk about each point in the memo. This would be the skill of finding and using supporting details.

3. Yes. I have had to read a floor plan of our office. It shows where the exits are on our floor so we know how to leave quickly in case there is a fire. It also shows where the offices and bathrooms are.

UNIT TWO: JOBS IN COMPUTERS AND ELECTRONICS

Lesson 4: Shipping Technology Products

CHECK YOUR UNDERSTANDING

page 31

Your answers may be similar to these:

1. Clue 1: You can combine customer orders going to the same customer at the same location.

Clue 2: You can combine later orders with earlier orders.

2. Clue 1: Ableset is a new customer.

Clue 2: Ableset's combined orders are over $50,000.

3. Clue 1: Combine orders to get better shipping prices.

Clue 2: Use reliable shippers that give reduced prices.

4. Clue 1: Big orders from new customers get shipped first.

Clue 2: New customers are called to make sure orders arrived in good shape and on time.

5. DeGear's order

Even though DeGear is a new customer, their order will be shipped last. DeGear's orders can't be combined because they are going to different locations.

6. Ableset is the only customer to have its orders combined. The orders can be combined because they are both going to the same location.

CHECK YOUR UNDERSTANDING
page 34

1. d
2. a and c
3. b and c
4. Your answer may be similar to this:

Nel decided that she must take care of the customer's problem right away. She called the packaging specialist in the shipping department to get help in correcting the problem.

5. If I were an employer, I would like to have Michael working at my company. He seems very responsible. Michael knew that Nel's customer was upset. Michael knew that he should solve the problem and give Nel a deadline for receiving the missing connectors.

LESSON WRAP-UP
page 35

Your answers may be similar to these:

1. I would infer that my supervisor liked my work. If she likes my work, she will recommend me for a raise.

2. I recently read a memo at work. The memo listed many problems in the customer service department. The memo did not blame anybody for the problems. However, I was able to infer from the memo that the supervisor had not given his workers the most up-to-date information.

Lesson 5: Building Electronics

CHECK YOUR UNDERSTANDING
page 39

1. b
2. Your answer may be similar to this:

The problem is that U17 and U18 logic chips face in the wrong direction. Henry should fix this problem by making them face in the opposite direction. The little circles need to be in the bottom right-hand corners.

3. a
4. Your answer may be similar to this:

There problems are that chips U3 and U2 are in the wrong spots. U7 and U8 are also in the wrong spots Henry needs to switch U3 with U2 and U8 with U7.

5. c
6. Your answer may be similar to this:

There are too many R11 and R12 resistors. There should only be one of each. The board has two of each. Henry should not put only one R11 and one R12 resistors above the U18 logic chip.

CHECK YOUR UNDERSTANDING
page 42

1. incorrect conclusion

Your answer may be similar to this:
Some of the company's benefits are good. Health, dental, and flexible schedules are good benefits. However, the employees want paid education benefits, too. They believe that the company should pay for outside work-related classes.

2. correct conclusion

Your answer may be similar to this:

Almost all workers saw some communication problems. Workers felt they had to attend too many meetings. Yet, they wanted some time to communicate with other employees. Most employees felt that communication needed to improve.

3. Your answer may be similar to this: Reducethe number of formal meetings and use other ways of communicating. Supervisors and team leaders could meet with teams over lunch to share ideas. Workers could also begin a newsletter. Informal get togethers could be held for special occasions, such as employee birthdays.

LESSON WRAP-UP

page 43

Your answers may be similar to these:

1. I think "jumping to the wrong conclusion" means that you don't get your reasoning right. You make the wrong judgment.

2. I think that Richard just doesn't care enough about his job to get here on time. I conclude that he'll probably be fired.

3. After I look more closely, I think that he is hurt. Maybe he was in an accident. Maybe that's why he's late. If this is the case, he probably won't get fired.

4. Your answer may be something like this:

One time I jumped to the conclusion that my sister was mad at me for no good reason. Actually, she was just in a bad mood and cranky because she had to stay 45 minutes late at work. So when she got home, she snapped at me every time I asked her a question. I jumped to the conclusion that she was mad at me, but she wasn't.

Lesson 6: Desktop Publishing

CHECK YOUR UNDERSTANDING

page 47

1. fact
2. fact
3. opinion
4. opinion
5. fact

Your answers may be similar to these.

6. a. opinion

b. Serious fonts may be a bad choice some of the time, but not always. The writer says they are a bad choice for a carnival poster. That could be right, but that's an opinion.

7. a. opinion

b. Yes, computers can do amazing things. The writer is stating an opinion by using the word *magic* to tell what a computer can do.

8. a. fact

b. This statement could be proven true. In fact, the reading says that "research has shown that this [too many fonts] confuses the reader."

CHECK YOUR UNDERSTANDING

page 50

1. b
2. c
3. opinion
4. fact
5. opinion
6. opinion
7. fact

Your answers may be similar to these:

8. Fact: The booklet shows information about how people can work safely at the computer.

Opinion: Ben thought the booklet was great because it showed lots of ergonomics information.

LESSON WRAP-UP

page 51

Your answers may be similar to these:
Dear Supervisor:

I am sitting in a bad work spot. I am requesting you to move me to a better spot.

It is a bad spot because I am in the middle of two people who don't get along.

During the day, each person mumbles bad things about the other. This happens a lot. They also give each other stares. They are nasty stares. All this is making it hard for me to work.

Sincerely,
(Your name)

UNIT TWO REVIEW

page 52

Your answers may be similar to these:

1. It might be difficult to tell if he was telling me a fact or his own opinion. One way I could tell is to check with the supervisor. If my supervisor says the area must be clean, the statement would be a fact. If my supervisor says it doesn't need to be clean, the statement is his opinion.

2. If the comment was my co-worker's opinion, I would think he interferes too much. I would conclude he is a meddler. He is someone who sticks his nose in other people's business.

3. Yes. I have had to tell a fact from an opinion at work. Sometimes information that I get in a memo tells me that something is wrong. I then check with the person who wrote the memo to see if the information is really wrong or if it's his belief.

UNIT THREE: COMMUNICATIONS TECHNOLOGY OCCUPATIONS

Lesson 7: Working with Mobile Phones

CHECK YOUR UNDERSTANDING

page 57

1. a
2. c
3. b
4. c
5. a
6. b

CHECK YOUR UNDERSTANDING

page 60

1. c, d

2. d

3. Your answer may be similar to this:

It's good to have the theft alarm feature for a car phone because it can alert you if your car or your phone is being stolen. This feature doesn't really make sense for an ordinary house phone.

4. Your answer may be similar to this:

Super speed dialing is a good feature for both types of phones. It saves time for the person dialing. It's a convenience. Also, it lets car phone users keep their eyes on the road more. They need less time for dialing.

5. Your answer may be similar to this:

Automatic redial is a good feature for a car phone user. It lets the driver focus on the road. The driver doesn't have to keep pressing every key to redial the number. The phone does it for the driver.

LESSON WRAP-UP

page 61

Your answers may be similar to these:

1. Since he has installed phones in the same kind of car, he might compare this customer's car with a previous car. He might see that they have the same inside structure. The cars might also have the same places for installing the equipment.

2. The MCT might contrast the shape and size of this customer's phone to those he has installed before. If the phone is bigger or smaller, he would need to make some adjustments. The MCT might also contrast all the pieces to be installed with this phone against the pieces used in other installations.

Lesson 8: Developing Film

CHECK YOUR UNDERSTANDING

page 65

1. c
2. b
3. d

Your answers may be similar to these:

4. A shiny surface can cause glare spots in flash photos.

5.

The photographer can:

- Take the picture at an angle to the subject, rather than directly in front of the shiny surface.

CHECK YOUR UNDERSTANDING

page 68

1. a
2. d
3. a, c
4. d
5. Your answer may be similar to this:

PLTs must keep the equipment and the darkroom clean so that dust and other particles don't get on the film or print paper. If they do, they can cause spots in the film and on the printed photo. The customer won't want to see these effects on the film or prints.

LESSON WRAP-UP

page 69

Your answers may be similar to these:

1. There is a lot of information on a warning label that would be helpful. You should know what chemicals are in the bottle. You should know the kinds of problems these chemicals can cause. Many warning labels also include directions on how to treat a problem if it occurs. This information is also good to know.

2. Strong cause-and-effect reading skills are important for trying to find and solve problems on the job.

Lesson 9: Installing Satellite Dishes

CHECK YOUR UNDERSTANDING

page 73

1. c
2. d
3. b
4. a

Your answer may be similar to this:

5. It is important that the bolts are secure so the dish doesn't move when it isn't supposed to. The dish must face the satellite directly. If the bolts aren't tight enough, the dish won't hold its position.

CHECK YOUR UNDERSTANDING

page 76

1. d
2. a
3. c
4. a
5. b
6. d
7. Your answer may be similar to this:

Extreme temperatures can crack the cable's casings. Snow and ice can affect reception by weighing dish down.

LESSON WRAP-UP

page 77

Your answers may be similar to these:

1. Installers must be able to follow directions when they set up a satellite dish system. For example, they use this skill to find the right spot for the dish. They use this skill to put together and mount the dish. Installers also use this skill when they show customers how to use and care for their systems.

2. Installers must be able to follow directions to get the system working right. For example, they need to follow directions to find out why reception is bad.

UNIT THREE REVIEW

page 78

Your answers may be similar to these:

1. I would use the skill of identifying cause and effect to explain what went wrong with the project. For example, I could tell her that not buying the right part caused a delay in the project.

When I tell her in a memo how I would fix the problem, I would most likely use the skill of comparing and contrasting. I would compare the original schedule with the new schedule and then draw the conclusion that I could make up the time.

2. I would probably use the skills of comparing and contrasting a lot when reading his letter. He probably will tell me why one computer is better than another. He probably has compared and contrasted

lots of their features.

I also would use the skill of following directions when he tells me where to buy the computer.

3. Yes. I need to follow directions at work a lot. Sometimes, I mix chemicals. I check the label for the right amounts. I also follow the order of the steps in the directions. Sometimes the directions also have a diagram. The diagram helps me follow the steps

UNIT FOUR: ENVIRONMENTAL TECHNOLOGY OCCUPATIONS

Lesson 10 Working in Air Pollution Control

CHECK YOUR UNDERSTANDING

page 83

1. A half-mask, air-purifying respirator is the simplest respirator that can be used on an asbestos job. It is the first sentence in the first paragraph.

2. They explain which part of the mask goes under your chin and which part goes over your nose.

3. The half-mask will not work unless the filter is made especially for asbestos .

4. A respirator's protection factor tells how many fibers can leak inside

5. You need to check your respirator before you put it on.

6. Your answer may be similar to this:

I know that the main idea of each paragraph can be a detail supporting the main idea of the whole passage. I looked at the stated or unstated main idea of each paragraph. I put them together and thought about what they had in common (as details). Then, I used these details to find the main idea, or most important point, of the passage. It is A half-mask respirator is the simplest respirator that can be used on an asbestos job.

7. Some possible answers are:

How to Use a Half-Mask, Air-Purifying

Respirator

A Proper Mask to Wear for Asbestos Removal

Selecting a Mask to Wear for Asbestos Removal

CHECK YOUR UNDERSTANDING

page 86

1. b

2. c

3. a, c, d (b is the main idea)

4. b

5. Your summary may look something like this:

There are plenty things you can do to prevent heat stress and heat stroke. Drink plenty of water. Eat foods that help your body hold water. Take breaks. Allow your body time to get used to the heat. Avoid alcohol. Watch out for signs of heat stress or heat stroke.

LESSON WRAP-UP

page 87

Your answers may be similar to these:

1. a. the title

   b. the first headline

   c. the first paragraph

2. how to find the main idea in the title, in a headline, or in the first paragraph; how to look for main ideas in the paragraphs of a longer passage; and how supporting details explain a main idea and give me information I may need to remember.

Lesson 11: Working in Waste Management

CHECK YOUR UNDERSTANDING

page 90

1. T

2. F

3. N

4. T

5. F

6. T

7. N

8. N

9. F

CHECK YOUR UNDERSTANDING
page 93
   1.c, d
   2.b, c
   3.d
   4.a
   5.a, c
   6.toxic material that is a health hazard
   7.ctoxic material that may catch fire
   8.no hazardous material

LESSON WRAP-UP
page 94
Your answers may be similar to these:
   1.a. Without my protective clothes, I may place myself in danger, so I will not take the job.
      b. I don't have my protective work clothes.
I can't get them out of the locker room.
My friend didn't say the company would provide clothes.
He said I may only need gloves, and he doesn't think the material is contaminated.
But that is his opinion.
I have been trained not to take chances.
2.

Reasons	Solutions
Not able to understand it.	Ask for help.
Read too fast.	Read slowly; read twice.
Information is missing.	Ask for information needed.
Wrong information given.	Use past experience to decide if information can be trusted.

Lesson 12: Working in Recycling Jobs

CHECK YOUR UNDERSTANDING
page 99
   1.Your list should look similar to this:
Our Plant
1. plain white paper
2. Phone book—White Pages
3. fax machine paper
North Shore Plant
4. plain colored paper
5. colored folders
6. newspapers
7. Phone book—Yellow Pages
8. magazines
9. glossy brochures
Can't Use
10. junk mail letters
11. envelopes with labels
12. envelopes without labels
13. sticky notes
14. paper with labels
   2.Your list should look similar to this:
Small
1. envelopes with labels
2. envelopes without labels
3. sticky notes
Medium
4. plain paper—white
5. plain paper—colored
6. colored folders
7. junk mail letters
8. glossy brochures
9. fax machine paper
10. paper with labels
Large
11. Yellow Pages
12. White Pages
13. newspapers
14. magazines

CHECK YOUR UNDERSTANDING
page 102
Points for Using Paper
   1. renewable resource
   2. does not kill forests; comes mostly from tree farms
   3. often made from recycled fibers
   4. can easily be recycled
   5. accepted by many communities for recycling
   6. big paper bag recycling industry
Points Against Using Paper
   1. does not last long
   2. does not degrade any faster than plastic
   3. weighs more than plastic
   4. bulky to store
Points for Using Plastic

1. can be reused
2. lasts long
3. weighs less than paper
4. not bulky to store
Points Against Using Plastic
1. does not degrade fast
2. uses oil, a non-renewable resource
3. bad for fish and other forms of life

LESSON WRAP-UP
page103

Resist	Reduce	Reuse
c	a	b
f	d	e
g		h

UNIT FOUR REVIEW
page 104

Your answers may be similar to these:

1. First, look to see if the main idea is stated in the title or in a heading at the beginning of the selection. Then, look for the main idea in the first paragraph. Also look at the main idea of each paragraph. These main ideas may be supporting details for the main idea of the entire passage.

2. Read the information.

Look for main ideas.

Pay attention to details.

Connect the information to what you already know.

Make judgments or decisions based on the material.

Test your conclusions by asking yourself questions.

3. You can see related pieces of information. This helps you understand it more clearly. Classifying also helps you when you need to use information to make decisions. It makes it easier to make comparisons.

4. Being able to find the main idea will help you understand information on the job. As you read the main ideas of paragraphs and supporting details, you can connect them to the main idea of the whole piece. Finding the main idea helps you remember details that you might need in your work.

Knowing how to draw conclusions from what you read helps you avoid making mistakes, which is important on the job. When you think through information, you are able to make judgments based on the facts. Sometimes, you make have to make a quick decision when dealing with managers or customers. Knowing how to draw conclusions from information helps you "think on your feet."

When you classify information, you break it down into groups. These groupings help you see the information more clearly. Classifying also helps when you want to use the information to make decisions. When you classify information, you can use whatever grouping makes sense. It depends on how you want to use the information in your work.